Identification of Cleaner Production Improvement Opportunities

Identification of Cleaner Production Improvement Opportunities

Kenneth L. Mulholland
Wilmington, Delaware

AIChE®

WILEY-INTERSCIENCE

A JOHN WILEY & SONS, INC., PUBLICATION

A joint publication of the American Institute of Chemical Engineers and John Wiley & Sons, Inc.

Published by John Wiley & Sons, Inc., Hoboken, New Jersey.
Published simultaneously in Canada.

For general information on our other products and services or for technical support, please contact our Customer Care Department within the United States at (800) 762-2974, outside the United States at (317) 572-3993 or fax (317) 572-4002.

Wiley also publishes its books in a variety of electronic formats. Some content that appears in print may not be available in electronic format. For information about Wiley products, visit our web site at www.wiley.com.

Library of Congress Cataloging-in-Publication Data:

Mulholland, Kenneth L., 1939–
  Identification of cleaner production improvement opportunities / Kenneth L. Mulholland.
      p. cm.
  ISBN-13 978-0-471-79440-0 (alk. paper)
  ISBN-10 0-471-79440-6 (alk. paper)
  1. Chemical industry—Waste minimization.   2. Manufacturing processes—Production control.   3. Green technology.   4. Pollution prevention.
  I. Title.
  TD899.C5M85   2006
  658.5'67—dc22   2005030393

Printed in the United States of America.

10 9 8 7 6 5 4 3 2 1

Contents

Acknowledgments

The author wishes to acknowledge the valuable contribution of a number of colleagues at The Dow Chemical Company, Korean Institute of Industrial Technology and Hanwha Chemical. Specifically:

Duane Koch of The Dow Chemical Company for providing me with the opportunity to develop the manual.

Ray L. Schuette of The Dow Chemical Company for helping me with the initial development and implementation of the manual.

Dr. Kwiho Lee of the Korea Institute of Industrial Technology for working with me to further prove the effectiveness and value of the manual.

Mr. K. H. Hyun of the Hanwha Chemical Corporation for providing the resources and guidance while implementing the cleaner production technologies at the Ulsan vinyl chloride monomer plant.

Foreword

One of the very first-pollution prevention projects I recall was such a "no-brainer" that any first year chemical engineering student would have spotted it in a heart beat. in this case, a Fortune 100 company was taking a concentrated waste stream, mixing it with a largely aqueous stream and sending it off for treatment. In other words, a low-volume, highly-contaminated stream had been transformed into a large volume waste stream with low contaminant concentration. It doesn't take much of an imagination to realize that the treatment cost of the latter scheme would be enormously higher than the cost of treating the more concentrated, low-volume stream.

Why did the above happen? I like the story that the author of this manual, Dr. Ken Mulholland tells in a June, 2003 paper that he published in Chemical Engineering progress entitled "Think Outside the Box to Reduce Wastes." To quote Ken, "Consider the person who walks along a circular path in the same direction every day. The first time Every thing is new. By the 30th time, the walker only notices the unusual. The slow deterioration of the walkway or cumulative minor changes in the landscape are not evident. However, if on the thirty-first walk, the person walks in the opposite direction, all of a sudden everything is new, It is the same scenery, but now it is seen from a different perspective."

"The same holds true for a manufacturing process," according to Ken. "Instead of starting from the front of the process, if you start with the waste streams and move backward through the process, and ask different questions of the same process information, your views change completely."

All well and good you say, but so what? Why should I be bothered to spend money in this manner to identify process improvements that might reduce my wastes? Dr. Paul Tebo of DuPont said it best several years ago: if we use pollution prevention as a weapon, we can increase yield, eliminate the need for some of our expansion projects, lower our treatment costs, reduce permitting requirements and thereby gain a competitive edge on our competition. In other worlds, it is all about money. Many would argue (I amongst them) that such thinking leads one on a pathway to sustainability and improved community relations.

The example I cited above is clearly, in the vernacular, low hanging fruit. How does one get at the more intractable problems that are more process driven? As an example, how did the waste in that original low flow highly contaminated waste stream get there in the first place? And how do I eliminate them?

In this era where we 'talk the talk" about sustainability, Ken provides the tools one needs to actually "walk the walk" to "meet the needs of the present without compromising the ability of future generations to meet their own needs." *

* Atkisson, Alan. 1999. *Believing Cassandra, An Optimist Looks at a Pessimist's World.* White River Junction: Chelsea Green Publishing Company.

Richard D. Siegel, Ph.D.

Director, Industrial Services & Principal Consultant

K. M. CHNO ENVIRONMENTAL, INC. Burlington, MA

Preface

Processes that produce waste reduce profitability. Pollution prevention, waste minimization, and cleaner production programs can reduce waste generation by 40–50% with a 200% internal rate of return. Unfortunately the education and training of chemical engineers have resulted into a thought framework or "box of thought" leading to process designs that produce excessive waste and thus have higher investment and operating costs. The engineer needs to be shown how to think outside the "box" and discover process improvements that traditional engineering problem-solving techniques do not find.

Consider the person who walks along a circular path in the same direction every day. The first time everything is new. By the 30th time, the walker notices only the unusual. The slow deterioration of the walkway or cumulative minor changes in the landscape are not evident. However, if on the 31st walk the person walks in the opposite direction, all of a sudden everything is new. It is still the same scenery, but now it is seen from a different perspective.

The same holds true for a manufacturing process. Instead of starting from the front of the process, if you start with the waste streams and move backward through the process, and ask different questions of the same process information, your view changes completely.

The basic skills and knowledge of the engineers, operators, mechanics, scientists and business people are available to solve waste generation problems. The information is available. All that is needed is a different approach to looking at the information

The cleaner production/pollution prevention technologies were described in the book *Pollution Prevention Methodology, Technologies and Practices* that was published by AIChE in 1999. Using the book's information and technology, I worked with The Dow Chemical Company to develop this manual. I further refined the manual while working with the Korean Institute of Industrial Technology.

To accelerate the introduction of cleaner production technologies to various industries throughout South Korea, in 1999 the Ministry of Commerce, Industry and Energy formed the Korea National Cleaner Production Center (KNCPC) as a division of the Korea Institute of Industrial Technology. In 2001 I worked with KNCPC to develop cleaner production technology for complex processes. Hanwha Chemical Corp. volunteered to provide a process and resources to demonstrate the technology.

Hanwha Chemical manufactures polyvinyl chloride (PVC), low-density polyethylene (LDPE), linear low-density polyethylene (LLDPE), and chloralkali. It is headquartered in Seoul, has a research and development center in Daejeon, and operates two manufacturing plants — Yosu and Ulsan. Hanwha's policy is to follow stricter environmental management guidelines than required by regulation, and it has adopted the ISO 14001 Environment Management System. Even though Hanwha had just completed a massive energy conservation program, the company's leadership understood the value of cleaner production technologies and volunteered to work with KNCPC to develop such technologies for complex processes.

The project focused on the Ulsan vinyl chloride monomer plant, since it had a higher manufacturing cost and environmental load. Plant personnel were trained on how to view their process by focusing on the waste streams instead of the product streams. A brainstorming team, consisting of experts in chemistry, engineering, environmental control, electricity, piping and equipment, and operations, met and generated more than a hundred process improvement ideas. The top ideas underwent further technical and economic analyses, and an implementation plan was formulated for the best ideas. The process improvements required no new technology, only better use and understanding of existing technologies, and ranged from revised operating procedures to major process modifications that can be patented. The improvements, which are expected to be completed by the end of 2003, involve a capital investment of $2,600,000 and will achieve:

- a 35.7%, or 5,232-m.t., waste reduction

- a cost reduction of $3,200,000

- additional revenue generation of $6,000,000.

Hanwha Chemical is expanding the program to all of its processes at the Ulsan site. KNCPC is actively implementing and promoting cleaner production technologies for all industries. It is working to: develop metrics by industry type; disseminate technology through collaborations with universities, institutes and research teams, as well as via forums such as roundtables; and provide support for businesses developing and implementing cleaner production technologies.

In summary this book contains a tested technology that will lead your business personnel to discover process improvements that will reduce your waste generation, reduce the resources requirements to manufacture your product and increase the revenue to your business.

Section I

Cleaner Production and Waste

Introduction

Processes that produce waste reduce profitability. Pollution prevention, waste minimization, and cleaner production programs can reduce waste generation by 40–50% with a 200% internal rate of return. [1,2] Unfortunately the education and training of chemical engineers have resulted in a thought framework, or "box of thought", that leads to process designs that produce excessive waste and thus have higher investment and operating costs. The engineer needs to be shown how to think outside the "box" and discover process improvements that traditional engineering problem-solving techniques do not find. (Figure I-1)

Consider the person who walks along a circular path in the same direction every day. The first time everything is new. By the thirtieth time, the walker only notices the unusual. The slow deterioration of the walkway or cumulative minor changes in the landscape are not evident. However, if on the thirty-first walk the person walks in the opposite direction, all of a sudden everything is new. It is still the same scenery, but now it is seen from a different perspective.

The same holds true for a manufacturing process. Instead of starting from the front of the process, if you start with the waste streams and move backward through the process, and ask different questions of the same process information, then your view changes completely.

The basic skills and knowledge of the engineers, operators, mechanics, scientists and business people are available to solve waste generation problems. The information is available. All that is needed is a different approach to looking at the information.

The book *Pollution Prevention Methodology, Technologies and Practices* [3] contains more detail on the cleaner production/pollution prevention technologies described in this manual. This manual's information and technology were tested at The Dow Chemical Company and further refined with the Korean Institute of Industrial Technology.

In summary, this manual contains a tested technology that will lead business and design personnel to discover process improvements that will reduce waste generation, reduce resource requirements to manufacture product(s) and increase revenues to the business.

1

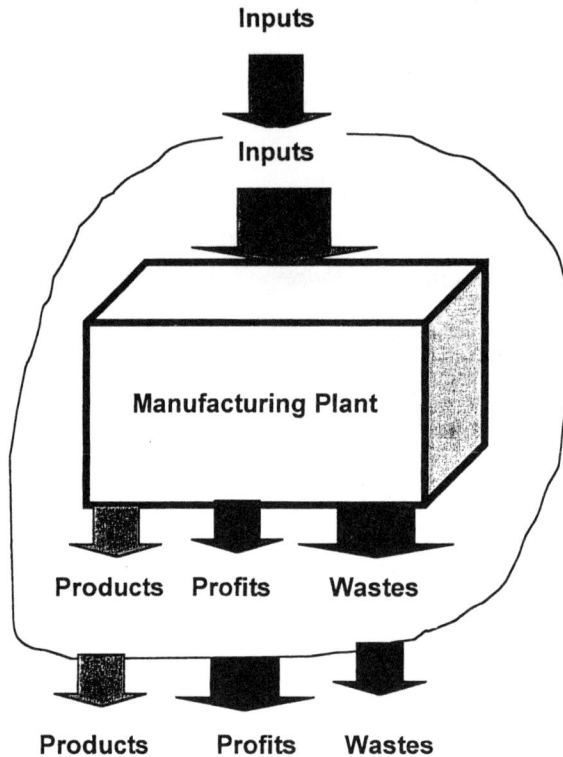

Fig. I-1. Think outside the "box", reduce inputs and wastes and increase profits

Manual

The manual consists of five (5) sections:

- Section I discusses the relationship between cleaner production, waste, cleaner production analyses and the business value derived by using a cleaner production methodology to identify process improvement opportunities.

- Section II covers how to develop a cleaner production program and to identify which businesses and waste streams to attack first.

- Section III discusses the data requirements and data analyses to prepare for the opportunity identification step.

- Section IV is the opportunity identification step.

- Section V covers how to assess and rank the best ideas.

- Appendix A contains copies of the forms and handouts needed to accomplish different steps in the methodology.

- Appendix B describes a chemical process case study to illustrate the methodology.

Cleaner Production and Sustainable Manufacturing

The traditional approach to process design is to first engineer the process and then to engineer the treatment and disposal of waste streams. However, with increasing regulatory and societal pressures to eliminate emissions to the environment, disposal and treatment costs have escalated exponentially. As a result, capital investment and operating costs for disposal and treatment have become a larger fraction of the total cost of any manufacturing process. For this reason, the total system must now be analyzed simultaneously (process plus treatment) to find the minimum economic option.

Experience in all industries teaches that processes that minimize waste generation at the source are the most economical. For existing plants, the problem is even more acute. Even so, experience has shown that waste generation in existing facilities can be significantly reduced (greater than 30% on average), while at the same time reducing operating costs and new capital investment. Also, experience has shown that processes, which generate waste, require 10% to 35% more investment. The higher investment is required to store, heat, move and separate the waste streams from the product(s) and recyclable feed materials, solvents, catalysts, and so on.

Cleaner production technology provides tools, which address the problems of negative environmental impact, loss of materials to waste, and increased process investment to deal with wastes. The application of these tools will result in a manufacturing process evolving from a typical process shown in Figure I-2 to the desired process shown in Figure I-3.

The most common definition for sustainable development is development that "meets the needs of the present without compromising the ability of future generations to meet their own needs."[4] For a manufacturing facility the definition translates to a process that requires the minimum amount of resources and produces no waste, that is, Figure I-3.

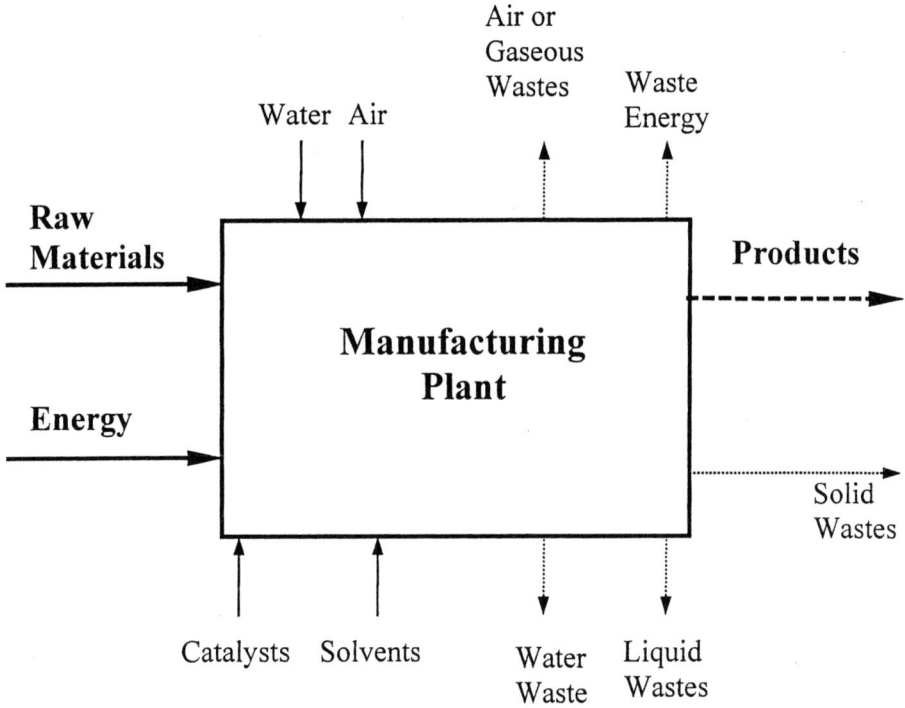

Figure I-2. Plant with Pollution.

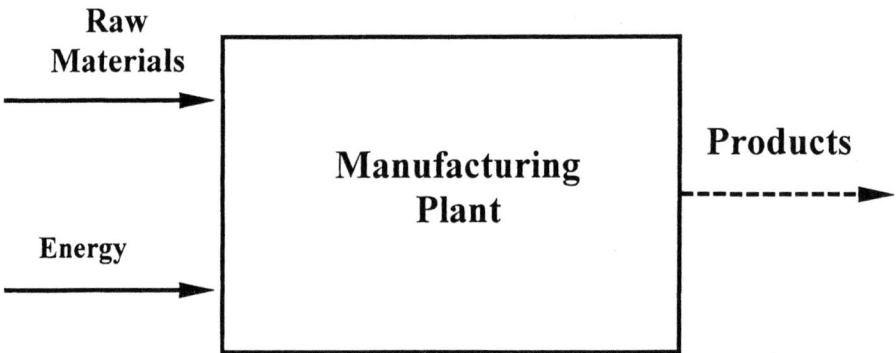

Figure I-3. "ZERO" Waste Generation Facility

The concepts behind the methodologies presented in this manual are described in the book "Pollution Prevention: Methodologies, Technologies and Practices." The last chapters of that book describe technologies and practices to reduce the waste from processes ranging from ventilation problems to reactors to pH control.[3]

Why Waste?

The standard college education, normal process design procedures and on-the-job training are three contributing factors as to why manufacturing processes produce excessive waste (Figure I-2). The education, design and training focus has been on the product.

College Education

Until recently, for the last 5 to 10 years, when the vast majority of plants now operating within the U.S. were designed, the teaching of process design focused on:

1) Lowest possible investment,

2) Lowest operating cost, and

3) Highest throughput possible, especially first pass yield.

These three major objectives of process design were developed in an environment which now seems both innocent and naïve because of the beliefs that:

1) Air was an infinite sink - that is the gases disappeared and were not even considered as harmful or at worst were easy to treat,

2) The lines that left the flow-sheet for treatment were someone else's problem, and

3) Energy costs were not significant.

Further exacerbating the problem was the fact that society was not aware of the problems being caused by waste. Jobs were more important than any harm to the environment or discomfort to the workers, for example, coal miners and coal dust, mill workers and ergonomics, smelter workers and fumes, farm workers and pesticides and so on.

Process Design

The normal design process has the following steps.

5

1) Researchers develop a new product.

2) Engineers synthesize a design using their previous experiences around the criteria of minimum investment, minimum operating cost and maximum throughput.

3) The design is further refined to make maximum profit.

4) Just before final design any necessary treatment facilities to meet regulatory requirements is then considered, almost as an afterthought.

5) Construction and finally startup.

Once the plant is running the businessperson directs the plant manager to run the process to offset startup costs and generate revenue. The plant manager cannot take time to optimize the process - the business needs the cash.

Training of New Engineers

Now comes the training of the engineers to solve process problems. An experienced engineer mentors an engineer new to the process. The mentor starts with the feed system and discusses the problems associated with the feed system, safety etc. The mentor explains how the reactor operates, temperatures, pressures, catalysts, hazards, etc. Then the mentor describes the separation systems, packaging and shipping. Finally the mentor states that the process does generate some waste and this or that is the waste treatment system. This system is described almost as an afterthought.

The "new" engineer is asked to solve some problem to see how well he/she will handle the challenge. Having no preconceived ideas the engineer uses his/her full range of skills and does a good job. Now it is 10 years down the road. The now "experienced" engineer solves the problems tomorrow the same way as the problems were solved yesterday. Over time while solving problems the engineer acquires blinders on all other happenings, almost as if they were at the Colorado River of the Grand Canyon instead of being on the upper plateau and being able to have a greater view of everything.

Result

The engineer's education, the design process and engineer's training at their job result in a thought framework or "box of thought" that result in process that produce excessive waste, thus require higher operating costs (Figure I-1). The need is to show the engineer how to think outside the "box" and discover process improvements that standard problem solving techniques do not find.

Cleaner Production Analysis Defines Improvement Opportunities
U.S. EPA and DuPont Chambers Works Waste Minimization Project[1]

In May 1993, the U.S. EPA and DuPont completed a joint two-year project to identify waste reduction options at the DuPont Chambers Works site in Deepwater, New Jersey.[1] The project had three primary goals as conceived:

1. Identify methods for the actual reduction or prevention of pollution for specific chemical processes at the Chambers Works site.

2. Generate useful technical information about methodologies and technologies for reducing pollution which could help the U.S. EPA assist other companies implementing pollution prevention/waste minimization programs.

3. Evaluate and identify potentially useful refinements to the U.S. EPA and DuPont methodologies for analyzing and reducing pollution and/or waste generating activities.

The business leadership was initially reluctant to undertake the program, and was skeptical of the return to be gained when compared against the resources required. After completing a few of the projects, however, the business leadership realized that the methodology identified revenue-producing improvements with a minimum use of people resources and time, both of which were in short supply.

The pollution prevention program assessed 15 manufacturing processes and attained the following results:

- A 52% reduction in waste generation.

- Total capital investment of $6,335,000.

- Savings and earnings amounting to $14,900,000 per year.

  — Reduced treatment costs   11%

  — Recovered product   15%

  — *Process Improvements*   *74%*

11 of the 15 manufacturing processes identified waste reduction opportunities that would require less than $50,000 capital investment and could be completed within 6 months.

The key to the site's success was following a structured methodology throughout the project and allowing the process engineers' creative talents to shine through in a disciplined way.

Cleaner Production Pograms

The methodology described in this manual has been used to identify process improvement opportunities for more than 50 processes, which include:

- Pharmaceutical intermediates,

- Elastomer monomers and polymers,

- Polyester intermediates and polymers,

- Batch processes such as agricultural products and paints

- Chlorofluoro-hydrocarbon and chlorocarbon processes,

- Specialty chemicals and waste water treatment facilities.

Waste

A waste is an unwanted byproduct or damaged, defective, or superfluous material of a manufacturing process. A secondary source of waste generation is the excess energy required to process and to treat any waste that is generated.

Manufacturing processes produce three classes of waste:

- Process wastes are solid, liquid, vapor and excess-energy-generated wastes resulting from transforming the lower-value feed materials to a higher-value product(s).

- Utility wastes are solid, liquid, vapor and excess-energy wastes resulting from the utility systems that are needed to run the process, e.g., steam, electricity, water, compressed air, waste treatment, etc.

- Other wastes result from start-ups and shutdowns, housekeeping, maintenance, etc.

Process Waste

Process chemistry often produces waste. For example:

$$A + B + C \longrightarrow I1 + BP1$$

$$I1 + D \longrightarrow P1 + BP2 + BP3$$

Where:

- A, B, C and D are feed materials.

- I1 and P1 are an intermediate and product.

- BP1, BP2 and BP3 are byproducts that have no value and require treatment or landfill.

To reduce the amount of byproduct waste, an economic value must be found for the byproducts (they would then be products), or the process chemistry has to be changed, for example

$$A + E \longrightarrow P1 + P2$$

Experience has shown that the **most costly waste** is process waste. If a business has limited resources, then the reduction of process waste should be the top priority. Lower process-waste generation results in:

- A lower cost-of-manufacture per pound of product, e.g., higher conversion of feeds to product(s), lower consumption of energy to move, store and process wastes, and so on.

- Reduced investment requirements per pound of product, i.e., lower process investment is required to process waste, or existing capacity that was required to process the waste is now made available. The investment required to move, store, separate and treat waste ranges from 10% to 35% of the total plant investment. For example, if air is being used to remove a waste (contaminant) from a stream, the end-of-pipe treatment investment is a function of the airflow rate. If the contaminant does not need to be removed, then this investment is no longer required.

- Decreased cost of manufacture in other areas, e.g., lower solvent losses, reduce energy use, lower manpower requirements, lower testing requirements, and so on.

Utility Waste

The major waste streams from utility systems are water, air, nitrogen and energy. The utilities required are a strong function of the manufacturing process, that is, the amount of waste that the process generates and how inherently safe it is. Examples of excess waste or losses from utility systems are:

- Venting of steam, inefficient steam traps, and a low level of condensate return.

- Inefficient boiler operation and steam distribution system heat losses.

- Compressed air and nitrogen leaks.

- Inefficient chilled-water and brine systems.

- Excessive use of chemicals and energy for waste treatment systems.

- Waste treatment sludges, incinerator particulate and acid gas control system discharges, excess energy use in incinerators and thermal oxidizers, contaminated carbon, and so on.

- Insufficient or compromised equipment insulation.

Preventing Industrial Pollution at its Source
Michigan Source Reduction Initiative[2]

The Natural Resources Defense Council, Dow Chemical, and a group of community activists and environmentalists initiated a project to reduce waste and emissions of 26 priority chemicals at Dow's Midland site by 35% using only pollution prevention techniques.[2] The project exceeded its goals and reduced:

- Targeted emissions by 43%, from 1 million to 593,000 pounds and

- Targeted wastes by 37%, from 17.5 million to 11 million pounds.

The cost savings and process improvements were significant where the waste reductions will be paid for in less than one year, a 180% overall rate of return. For example, one project required $330,000 and will return $3,300,000 per year in raw material savings. The 17 projects revealed important insights into pollution prevention:

- Majority of projects required relatively small amounts of capital. Ten were $50,000 or less. Five required no capital at all.

- Some of the greatest reductions in waste cost the least amount of money. Good reduction opportunities were found in almost every production process even though Dow businesses doubted that they would be found.

- Opportunities were broadly available in the various businesses in the plant.

- Several projects focused on basic process changes, yet were implemented in a relatively short period of time.

Other Waste

As with utility waste, the quantity of other wastes, such as uptime losses, maintenance waste, and storage losses, is strongly dependent on the amount of process and utility wastes required to run the process.

Examples of other waste include maintenance materials such as rags, filter cartridges, oils and greases, worn-out parts, gaskets and so on. Uptime losses include lost product, solvents, and catalysts.

Pollution

Pollution is any release of waste to environment (i.e., any routine or accidental emission, effluent, spill, discharge, or disposal to the air, land or water) that contaminates or degrades the environment.

Thermodynamic fact—A compound introduced to or created in a manufacturing process will escape as a waste or will not be completely destroyed by an end-of-pipe treatment device. If the compound is flammable, toxic, bio-persistent or bioaccumulative, the health and safety of humans and the ecology of the environment will be adversely effected.

The goals of cleaner production programs are to:

- <u>Minimize Generation.</u> Reduce to a minimum the formation of nonsalable byproducts in chemical reaction steps and waste constituents, such as tars, fines, etc., in all chemical and physical separation steps.

- <u>Minimize Introduction.</u> Minimize the addition of materials to the process that pass through the system unreacted or that are transformed to make waste. This implies minimizing the introduction of materials that are not essential ingredients in making the final product. Examples of introducing nonessential ingredients include: (1) using water as a solvent when one of the reactants, intermediates, or products could serve the same function, and (2) adding large volumes of nitrogen gas because of the use of air as an oxygen source, heat sink, diluent or conveying gas.

- <u>Segregate and Reuse.</u> Avoid combining waste streams together without giving consideration to the impact on toxicity or the cost of treatment. For example, it may make sense to segregate a low-volume, high toxicity wastewater stream from several high-volume, low toxicity wastewater streams. Examine each waste stream at the source and determine which ones are candidates for reuse in the process or can be transformed or reclassified as a valuable co-product.

Cleanest Production – "ZERO" Waste

In contrast, Figure I-3 depicts a manufacturing facility with an absolute minimum or "zero" amount of waste being generated. Since most processes require some level of solvents, catalysts and other materials and normally produce byproducts in the reactor, to attain a "zero" waste process requires fundamental understandings of the process and how to implement changes inside the pipes and vessels.

"ZERO" Waste

Solvents used in the manufacture of intermediate monomers were incinerated as a hazardous waste. Alternative nonhazardous solvents were considered and rejected. However, the intermediate monomers were found to have the dissolution capacity of the original solvents and could replace them. By utilizing existing equipment, realizing savings in ingredients' recovery, and reducing operating and incineration costs, the project achieved a 33% internal rate of return (IRR) and a 100% reduction in the use of the original solvents.

Root Causes

There are two general categories of root causes for process waste—operational and fundamental. Operational root causes arise from how the process is operating versus how it should operate. Examples of operational root causes are:

- Poor understanding of the process,

- Control problems,

- Not following operating procedures, and

- Maintenance problems.

Process Understanding

At a DuPont site, tars were plugging a distillation column and feed preheater.[1] The tar buildup resulted in plant shutdowns every 3 months to clean the preheater and every 9 to 12 months to replace the column packing. This cost the business hundreds of thousands of pounds of lost production each year. A second preheater was installed in parallel with the existing heat exchanger to allow cleaning without shutting the plant down. In addition, lab and plant tests were run to better understand the mechanisms behind the tar formation.

As a result of these tests, the plant discovered that the tar formation reaction was pH-sensitive. By tightening Standard Operating Procedures and installing alarms on upstream crude product washers, the tar formation was virtually eliminated. This provided the business with approximately $280,000/yr in additional after-tax earnings.

Maintenance

Methylene chloride is used as a coating and cleaning solvent in the manufacture of several graphic arts and electronic photopolymer films.[3] In the late 1980s, a site released more than 3 million pounds of chlorinated solvents into the air. The coating solution preparation areas accounted for the majority of the air emissions. Coating solutions are prepared batchwise in agitated vessels using a blend of polymers, monomers, photoinitiators, pigments, and solvents. These atmospheric mix tanks were not well sealed, resulting in large fugitive and point source emissions of methylene chloride.

In an effort to enclose the batch vessels as much as possible, the mix tanks were fitted with bolted, gasketed lids. The vessels were also designed with pressure/vacuum conservation vents to allow the vapor pressure to rise to 3 psig before the tanks breathed. By sealing up the process, the site was able to reduce air emissions by 40% and save $426,000 per year in methylene chloride costs.

Fundamental root causes arise from the chemistry, thermodynamic and engineering limitations of the process. Examples of fundamental root causes are:

- Chemistry route picked that requires toxic solvents,

- Catalyst selection and byproduct formation,

- Reactor operation and byproduct formation,

- Not understanding the functions and thermodynamic principles of the separation processes, and

- Inadequate engineering of the process equipment.

Any unwanted material introduced to or created in a process will escape as a waste or will not be completely destroyed by an end-of-pipe treatment device.

Control of the Reaction Pathway

In hydrocarbon oxidation processes to produce alcohol, there is always a degree of over-oxidation.[3] The alcohol is often further oxidized to carboxylic acids and carbon oxides which are wastes. If boric acid is introduced to the reactor, the alcohol reacts to form a borate ester which protects the alcohol from further oxidation. The introduction of boric acid terminates the byproduct formation pathway and greatly increases the product yield. The borate ester of alcohol is then hydrolyzed, releasing boric acid for recycle back to the process. This kind of reaction pathway control has been applied to a commercial process, resulting in about a 50% reduction in waste generation once the process was optimized.

Thermodynamic Review

In distillation, the conventional wisdom is to remove the low-boiling material first.[3] After the low-boiling material was removed in a DuPont batch process, the remaining mixture was very difficult to separate, because of azeotropes and pinch points formed by the remaining compounds. The separation difficulties resulted in about one-third of the production run having to be incinerated.

The vapor-liquid equilibrium data for the compounds was reexamined—especially the binary interaction parameters. This reexamination revealed that the low-boiler could extract the product from the remaining compounds. A pilot plant test confirmed the concept, and a continuous extraction process was designed and constructed. The new process reduced the lost product from 200,000 lb./yr. to less than 2,000 lb./yr., and the impurities in the final product were decreased from 500 ppm by weight to less than 1 ppm by weight.

Opportunity Identification

Identification of opportunities to reduce waste from a manufacturing facility requires contributions from everyone, from process engineers and chemists, operators, mechanics, business people and so on. To identify the opportunities, a structured methodology is required that

- uses minimum resources (time and money) to define process improvement opportunities and to conduct a process baseline analysis,

- uses existing process information to define the process improvement opportunities,

- defines the process characteristics, i.e., process changes required inside the pipes and vessels to optimize the process, and

- has been proven to work.

The cleaner production methodology described in this manual meets all of these requirements. It has been used on over 50 processes and has in all cases identified new process improvement opportunities that minimize/reduce waste and in 75% of the cases identified improvements that cost less than $50,000.

The methodology is based partly on the observation that the volumetric flow of an air or gaseous waste stream and the volumetric flow and organic loading of a wastewater stream (Figure I-2) determine the required end-of-pipe treatment investment and operating cost. Manufacturing plant investment and cost of manufacture are also influenced by the same gaseous and water flows. A second observation is that end-of-pipe treatment is required only because the streams contain components that have to be abated or removed.

Waste Stream Analysis

The first step in implementing waste minimization is to identify the options that will eliminate or minimize the waste streams' volumetric flow (this has the greatest influence on investment and operating costs) or eliminate/minimize the components of concern (which are the reason to treat the stream). To uncover the best options, each waste stream should be analyzed as follows:

1) List all components in the waste stream, along with any key parameters. For instance, for a wastewater stream these could be water, organic compounds, inorganic compounds (both dissolved and suspended), pH, etc.

2) Identify the components triggering concern, *e.g.,* hazardous air pollutants (HAPs), carcinogenic compounds, wastes regulated under the Resource Conservation and Recovery Act (RCRA), etc.

- Determine the sources of these components within the process.

- Then develop process improvement options to reduce or eliminate their generation.

3) Identify the highest volume materials — often these are diluents, such as water, air, a carrier gas, or a solvent. These materials frequently control the investment and operating costs associated with end-of-pipe treatment of the waste streams and have a significant impact on the process cost of manufacture.

- Determine the sources of these high-volume materials within the process.

- Then develop process improvement options to reduce their volume or eliminate them.

4) If the components identified in Step 2 are successfully minimized or eliminated, identify the next set of components that have a large impact on investment and operating cost (or both) for end-of-pipe treatment. For example, if the aqueous waste stream was originally a hazardous waste and it was incinerated, eliminating the hazardous compound(s) may permit the stream to be sent to the wastewater treatment facility. However, this may overload the biochemical oxygen demand (BOD) capacity of the existing wastewater treatment facility, making it necessary to identify options to reduce organic load in the aqueous waste stream.

In a sold-out market situation, a DuPont intermediates process was operating at 56% of its peak capacity.[3] The major cause of the rate limitation was traced to poor decanter operation. The decanter recovered a catalyst, and fouling from catalyst solids caused its poor operation. Returning the process to high utility required a 20-day shutdown. During the shutdown, the vessel was pumped out and cleaned by water washing. The solids and hydrolyzed catalyst were then drummed and incinerated. A *waste stream analysis* identified three cost factors—the volume of wastewater that had to be treated, the cost of the lost catalyst, and the incineration cost.

An *analysis of the process* and its ingredients indicated that the decanter could be bypassed and the process run at a reduced rate, while the decanter was cleaned. A process ingredient was used to clean the decanter, enabling recovery of the catalyst ($200,000 per year value). The use of the process ingredient cut the cleaning time in half, and that, along with continued running of the process, eliminated the need to buy the intermediate on the open market. The results were a 100% elimination of a hazardous waste (125,000 gallons/yr.) and cash flow savings of $3,800,000 per year.

Waste Stream Analysis and Process Analysis

Process Analysis

The only materials that are truly valuable to the business are the raw materials for reaction, any intermediates, and the final products. Aside from the feed, other input streams (*e.g.,* catalysts, air, water, etc.) are required because of the designers' limited knowledge of how to manufacture the product without them. The function of these input streams is to transform feed materials into products. To reduce or eliminate them, the feed materials, intermediates, or products must serve the same function as those input streams, or the process needs to be modified to eliminate them.

For either a new or existing process, a process analysis consists of the following steps:

1) List all feed materials reacting to salable products, any intermediates, and all salable products. Call this List 1.

2) List all other materials in the process, such as non-salable byproducts, solvents, water, air, nitrogen, acids, bases, and so on. Call this List 2.

3) For each compound in List 2, ask "How can a material from List 1 be used to do the same function as the compound in List 2?" or "How can the process be modified to eliminate the need for the material in List 2?"

4) For those materials in List 2 that are the result of producing non-salable products (*i.e.*, waste byproducts), ask "How can the chemistry or process be modified to minimize or eliminate wastes (for example, 100% reaction selectivity to a desired product)?"

Sample questions

When coupled with the application of fundamental engineering and chemistry practices, examining a manufacturing process by these waste stream and process analyses' techniques will often result in a technology plan for a minimum-waste-generation process.

Typical questions are:

- If the solvent is "bad", can another more benign solvent be used?

- Can the process be modified to eliminate the solvent?

- If water is used for example to dissolve a salt of the intermediate, why use water? Why not transfer the intermediate as a solid?

- If multiple solvents are used, why?

- If a homogeneous catalyst is used and has to be separated, why not use a heterogeneous catalyst?

- If any reactions are energetic, is there a different reaction pathway?

- If air is being used as a source of oxygen, why?

- If some reactions are slow and some are fast, is the proper type of reactor being used? Fast reactions in a pipe reactor and slow in a batch reactor.

- If byproducts are produced, is there a reaction pathway or set of conditions that do not result in certain byproducts?

- If toxic or hazardous solvents or feed materials are required, can they be produced in-situ to eliminate shipping and storage?

- Do separations unit operations use no solvents, require minimum energy and produce minimum byproducts?

Resources and Duties

A single person could do the entire program. However, a team of business personnel is necessary to discover a larger number of improvement opportunities.

Person	Role
Business leader	Provides direction, resources and support
Process team leader	Leads the entire program
Team members	3 to 4 members comprising of process engineer, process chemist, lead operator/mechanic, and possibly an environmental specialist or project engineer.
Cleaner Production Expert	Advises the process team on the required information to assemble for the brainstorming and facilitates the brainstorming session.
Scribe(s)	Records ideas during brainstorming session.
Brainstorming Team	8 to 15 people. The process team plus other members of the business and outside experts.

Cleaner Production Expert

The cleaner production expert provides the guidance for an effective program. The major program objectives are to:

- Obtain the business leader's commitment to the cleaner production program. The business leader—

  - Provides resources to develop and implement the program,

  - Defines the objectives of the program,

  - Provides support for the business personnel involved in the program.

- Educate the core team leader on the cleaner production program.

  - Acquaint the leader with the sequence of events in the program,

  - Ensure the leader of support to make the program a success,

- Enlist the leader's active and enthusiastic support for the program.
- Educate the core team on the cleaner production program.
  - Acquaint them with the sequence of events,
  - Acquaint them with the data requirements,
  - Enlist their support for the program.
- Facilitate the brainstorming to identify improvement opportunities.
  - Acquaint everyone on their duties and responsibilities,
  - Ensure that an effective brainstorming session occurs,
  - Be a resource for the business leader and core team.

To fulfill the desired objectives the cleaner production expert should

1) Meet with the business leader.

2) Meet with the core team leader.

3) Provide for the core team a three-hour seminar/discussion session on the cleaner production program.

4) Become acquainted with the process by having the core team describe the process and be given a tour of the facilities.

5) Discuss data requirements and sources.

6) Develop a timeline for events.

Summary

The discovery of process improvements that reduce waste generation is a **human** problem **not** a technical problem. Our education, process design methodology and on-the-job training causes plants to be built and operated that generate billions of pounds of excess waste that is estimated to incur hundreds of billions of dollars of excess cost of manufacture not to mention threat of public relations debacles. One approach is for the engineer to think outside of the "box", that is, to look at the manufacturing process from the back to the front end of the process.

Proven success in South Korea

To accelerate the introduction of cleaner production technologies to various industries throughout South Korea, in 1999 the Ministry of Commerce, Industry and Energy formed the Korea National Cleaner Production Center (KNCPC) as a division of the Korea Institute of Industrial Technology. Hanwha Chemical Corp. recently worked with KNCPC and Kenneth Mulholland and Associates to develop cleaner production technology for complex processes.

Hanwha Chemical manufactures polyvinyl chloride (PVC), low-density polyethylene (LDPE), linear low-density polyethylene (LLDPE), and chloralkali. It is headquartered in Seoul, has a research and development center in Daejeon, and operates two manufacturing plants — Yosu and Ulsan. Hanwha's policy is to follow stricter environmental management guidelines than required by regulation, and it has adopted the ISO 14001 Environment Management Systemt. Even though Hanwha had just completed a massive energy conservation program, the companys's leadership understood the value of cleaner production technologies and volunteered to work with KNCPC to develop such technologies for complex processes.

The project focused on the Ulsan vinyl chloride monomer plant, since it had a higher manufacturing cost and environmental load. Plant personnel were trained on how to view their process by focusing on the waste streams instead of the product streams. A brainstorming team, consisting of experts in chemistry, engineering, environmental control, electricity, piping and equipment, and operations, met and generated more than a hundred process improvement ideas. The top ideas underwent further technical and economic analyses, and an implementation plan was formulated for the best ideas. The process improvements required no new technology, only better use and understanding of existing technologies, and ranged from revised operating procedures to major process modifications that can be patented. The improvements, which are expected to be completed by the end of 2003, involve a capital investment of $2,600,000 and will achieve:

- a 35.7%, or 5,232-m.t., waste reduction

- a cost reduction of $3,200,000

- additional revenue generation of $6,000,000.

Hanwha Chemical is expanding the program to all of its processes at the Ulsan site. KNCPC is actively implementing and promoting cleaner production technologies for all industries. It is working to: develop metrics by industry type; disseminate technology through collaborations with universities, institutes and research teams, as well as via forums such as roundtables; and provide support for businesses developing and implementing cleaner production technologies.

Kenneth L. Mulholland, Kenneth Mulholland and Associates

Kwiho Lee, Korea National Cleaner Production Center

K. H. Hyun, Hanwha Chemical Corp.

The EPA/DuPont[1], Michigan Source Reduction Initiative[2], KNCPC program, and other cleaner production projects showed that:

- Product changes are not required,

- New technology is not required,

- Any new investment has greater than 100% IRR,

- Waste generation reduced by 40% to 50%,

- The knowledge resides in the business,

- Everyone in the business can contribute, that is engineers, chemists, operators, mechanics, and so on, and

- The design and operation of low waste generation processes is a human not a technical problem.

References

1) U.S. Environmental Protection Agency, DuPont Chambers Works Waste Minimization Project, EPA/600/R-93/203, Washington, DC: U.S. EPA, Office of Research and Development, November 1993.

2) Natural Resources Defense Council and the Dow Chemical Company, Preventing Industrial Pollution at its Source, The Final Report of the Michigan Source Reduction Initiative, Dillon, CO: Meridian Institute, September 1999.

3) K.L. Mulholland and J. A. Dyer, Pollution Prevention: Methodology, Technologies and Practices, New York: AIChE, 1999.

4) A. AtKisson, Believing Cassandra, An Optimist Looks at a Pessimist's World. White River Junction: Chelsea Green Publishing Company, 1999.

Section II

Waste Stream Selection

Introduction

Chemical processes have multiple waste streams ranging from fugitive or equipment leaks to large volume gas and water streams, as many as 50 to 70 emissions points. Many businesses are unaware of not only the number of emission points, but also the waste rate from each point. Likewise most businesses are unaware of the costs of waste and emissions. The costs encompass not only end-of-pipe treatment, but also the cost of manufacture associated with

- raw material losses;

- heating, cooling, storing and moving of the waste materials;

- higher investment and operating costs to separate the product(s) from the wastes; and

- yield loss associated with waste generation.

This section describes methods to estimate the amount of waste being generated by a process, the costs associated with those wastes, and how to select the waste streams on which to identify improvement opportunities.

Process Waste

The business purpose of a manufacturing process is to take low cost feed materials and transform them into a high value product. The typical process shown in Figure II-1 (same as Figure I-2) shows that to transform low value feed materials into high value products many other materials are required such as solvents, catalysts, water, air, nitrogen, acids, bases, and so on. For a chemical process that involves a reaction such as:

$$A + B + C \longrightarrow I1 + BP1 \qquad\qquad (II\text{-}1)$$

$$I1 + D \longrightarrow P1 + BP2 + BP3 \qquad\qquad (II\text{-}2)$$

Where:

- A, B, C and D are feed materials.

- I1 and P1 are an intermediate and product.

- BP1, BP2 and BP3 are byproducts that have no value and require treatment or landfill.

The minimum amount of waste per unit mass of product is BP1 + BP2 + BP3. To reduce the amount of byproduct waste, an economic value must be found for the byproducts (they would then be products), or the process chemistry has to be changed, for example

$$A + E \longrightarrow P1 + P2$$

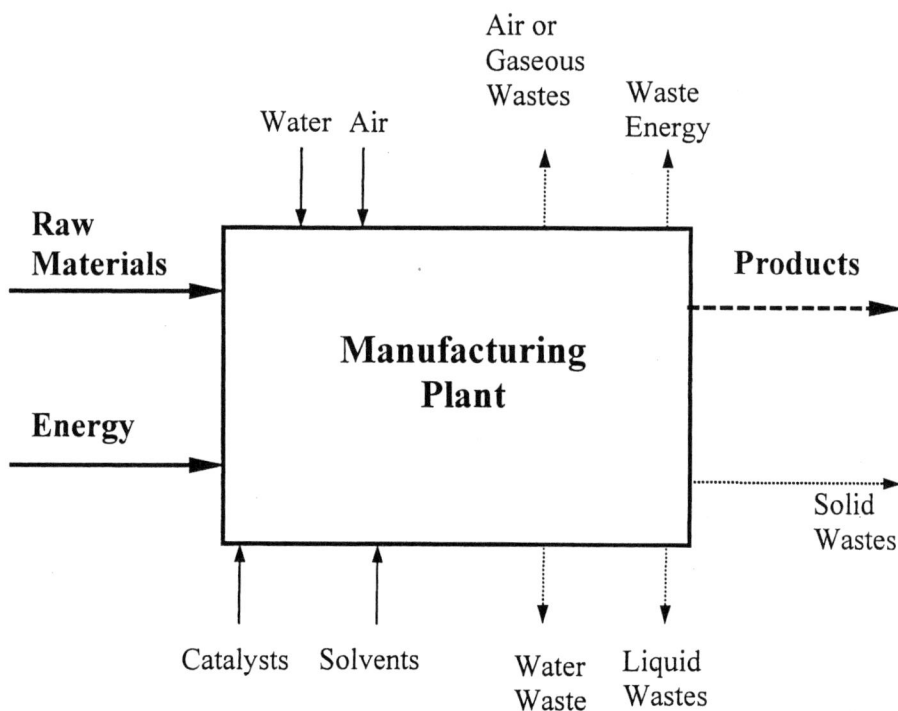

Figure II-1 Typical Manufacturing Plant

For the process shown in Figure II-1 and the chemistry in equations II-1 and II-2 the minimum waste is BP1 + BP2 + BP3. The actual waste generated by the Figure II-1 process will be many orders of magnitude higher. For purposes of this section waste is any material leaving a process that needs to be treated, for example water and air streams with contaminants and solid or other waste that has to be sequestered, and any untreated waste material such as fugitive emissions or equipment leaks.

To estimate the amount of waste for a given process:

1) Sum up all of the solvents, feed materials, catalysts, and so on that enter the process

- The data is available from chemical process mass balances as generated by ASPEN+ and PROII.

- Another source of data is the business' purchasing department.

2) Sum up all water (liquid and steam), air and other gases such as nitrogen and carbon dioxide that are introduced to the process—do not include utilities such as cooling water, heating steam or instrument air.

3) From the production data obtain the amount of product.

4) The amount of waste is the sum of **1)** and **2)** less the amount in **3).**

To determine the minimum amount of waste.

A) From the formulas in II-1 and II-2 calculate the mass of BP1 + BP2 + BP3 per mass of product P1.

B) Multiply the production rate by the minimum waste / mass of product.

Process Waste Calculation Example

The case study described in Appendix B will be used to illustrate the waste estimation technique described above. The process drawing is shown on page B-58 and the mass balances are listed on pages B-69 and B-70.

Process Waste

1) Material in feed Stream A1 plus 50 lb./hr benzene less recycle A24 = 49085 + 50 – 202 = 48933 lb./hr

2) Water and steam streams are A5 plus acid water plus steam for stripper = 17500 + 3000 = 20500 lb./hr

3) Product is stream A25 = 206 lb./hr

4) Waste = 49135 + 20500 – 206 = 69429 lb./hr

Minimum Waste

A) The process chemistry on page B-71 show 3 moles of water as waste per mole of product and the molecular weights on page B-73 result in 54 lb. waste / 99 lb. product or 0.545 lb. waste / lb. product.

B) At a production rate of 206 lb. / hr (stream A25), the minimum waste is 206 X 0.545 = 112 lb. /hr

Treatment Cost

The most obvious cost associated with waste streams is the end-of-pipe treatment costs. If no end-of-pipe treatment presently exists, then the installation of a new treatment device can be quite expensive. For existing end-of-pipe treatment devices the incremental cost for treatment is relatively small.

The end-of-pipe treatment investment and operating cost can be obtained by contacting a vendor, or the information in chapter 4 of the book *Pollution Prevention: Methodology, Technologies and Practices*[1] could be used to estimate the investment, operating cost and net present value for

- gas streams,

- aerosol containing gas streams,

- wastewater streams requiring pretreatment before being sent to the treatment system, and

- aerobic deep-tank activated sludge treatment facilities.

Treatment Cost Example

Streams A10 and A41 are the gas and liquid streams that require a new end-of-pipe treatment device. Stream A10 contains toxic compounds that need to be captured or destroyed. Stream A41 contains too high nitrogen content that needs to be removed before the stream is sent to the wastewater treatment facility.

Stream A10

Stream A10 has a 10,000 scfm flow rate. The graphs in the book *Pollution Prevention: Methodology, Technologies and Practices*[1] indicate that the investment would be $1,600,000 and the operating cost $560,000 / yr. (Figures 4-6 and 4-7).

Stream A41

Stream A41 has a flow rate of 35 gpm. The graphs in the book *Pollution Prevention: Methodology, Technologies and Practices*[1] indicate that the investment would be $800,000 and the operating cost $100,000 / yr. (Figures 4-20 and 4-21).

Value of Waste Minimization

Process investment and operating costs are governed by the types of materials added to or made in the process, and the amount of material sent through the process, for example 10,000 cu. M / min gas flows require more investment and operating costs than a 10 cu. M / min flow. To estimate the cost of waste:

I)   From the business' finance department obtain the total cost of manufacture for a given period.

II)   For that period estimate the total mass of material added to the process as was described above.

III)   Divide the total cost of manufacture by the total mass.

This calculation determines the cost per mass of material added to the process.

To estimate the value of waste reduction:

a. Subtract from the total waste calculated above in **4)** the minimum waste calculated in **B).**

b. Multiply that mass by the cost of manufacture / mass added to process.

Value of Waste Minimization Example

I)   The cost of manufacture = $ 624 / hr.

II)   Total mass to the process = 49135 + 20500 = 69635 lb. /hr

III)   On a per lb. of material into the process the cost of manufacture = $ 624 / 69635 = $ 0.009 / lb.

Value of Waste Reduction

a. Total waste less the minimum waste = 62429 − 112 = 62317 lb. /hr

b. The potential savings through waste reduction = 62317 X $0.009 = $ 560 / hr

Value of Improved Feed Material Utilization

For many processes there are significant quantities of feed materials that are lost either to the waste streams or are converted to waste byproducts. To estimate the lost product value:

1) Calculate the amount of product that would be produced from the limiting feed material per the reactions such as II-1 and II-2 on page II-1.

2) Subtract the actual production rate from the theoretical maximum.

3) Multiply the product price with calculated quantity in 2)

Value of Feed Material Utilization Example

The limiting feed material in the Appendix B case study is REAC.

1) The net feed to the process is REAC in stream A1 less the recycle in stream A24 = 483 −199 = 284 lb. / hr = 284 / 99 (Mol. Wt) = 2.866 moles / hr = 315 lb. / hr product theoretical maximum.

2) Lost production = 315 − 206 = 109 lb. /hr

3) Lost product value = $ 5.0 / lb. X 109 lb. /hr = $ 545 / hr.

Select Process Waste Streams

A typical process generates several major waste streams and many minor ones. Major can mean not only high volume or mass emissions, but also high toxicity, significant economic impact, etc. Selection of one or more of these major streams is needed for the first round of opportunity identification. If you are successful with these major streams, then you can target additional waste streams, including minor ones, in a second round. Notwithstanding the above, many minor streams offer opportunities for "quick hits" (i.e., projects that require little or not capital investment, offer a high probability of success, and are easy to implement). Even as you focus on the major waste streams, briefly visit the other waste streams to identify any "quick hits."

Identify Area Waste Streams

Remember that minor waste streams, such as leaks from pumps, fugitive emissions, and maintenance wastes, often provide "quick hit" opportunities. Do not spend time attempting to discover and document every possible fugitive emission or other non-significant waste stream from the process. Appendix A contains a waste stream description form. This form is used to document key information about each of the waste streams from the process—information that will be very valuable in identifying key waste stream variables to guide the identification of waste reduction opportunities. The waste stream description form contains key information about the source of the waste, how it is currently treated (if at all), composition and flow, and other measures, such as toxicity, odor, color, etc.

Waste Stream Description Form (Appendix A)

The information to be entered on this form is:

1) Date and identification information

2) Component information ordered from the highest to the lowest rate.

3) Component identification—by name, per cent in the stream, whether or not it is on any regulatory list and the source of the material.

4) Finally, input the stream total flow and any other information about the stream such as pH, odor or any other characteristics that would make it a target stream.

Prioritize Waste Streams

Business goals influence which waste streams are considered. The waste streams should be ranked based on one or more criteria. Some typical criteria or considerations include:

- If this waste stream is targeted, what is the likelihood of process improvements, such as increased yield, reduced cycle time, improved product quality, etc.?

- What is the economic incentive for addressing this waste in terms of avoided capital investment in end-of-pipe treatment and lower cost of manufacture?

- Is the waste disposed outside the plant boundaries in a landfill or in a commercial incinerator? Would the business prefer to increase control of its future by reducing its dependence on and the costs associated with outside resources to dispose of waste?

- What is the present or expected regulatory exposure to this waste stream?

- Is this a large volume stream in terms of flow?

- Does the waste contain compounds that are highly toxic? Are carcinogens or suspected carcinogens present?

- Does the waste stream present a special safety hazard, such as flammability, reactivity, or volatility?

- Does the waste contribute additional "soft-costs" to the business, such as lack of greenness, poor public image, and future liabilities?

If there are a large number of streams, then a Pareto chart can be used to provide visual guidance on which streams to select. A Pareto chart is a bar graph with the X axis being each waste stream and the Y axis having values such as stream total volumetric or mass flow, mass flow of listed components, or total mass of only the contaminants (See Figure II-2).

Select the Targeted Waste Streams

The number of waste streams selected will depend on the availability of resources. Choose a mix of major (at least one) and minor streams (such as "quick hits"). With the increased emphasis that has been placed on the environmental performance of a process over the last decade, the number of targeted waste streams will typically be 10 or less. The emphasis in this manual is on wastes that are made by the chemical process itself, such as unwanted byproducts, contaminated water and solvent streams, cleaning wastes, and volatile organic compounds (VOCs).

Figure II-2 – Pareto Chart

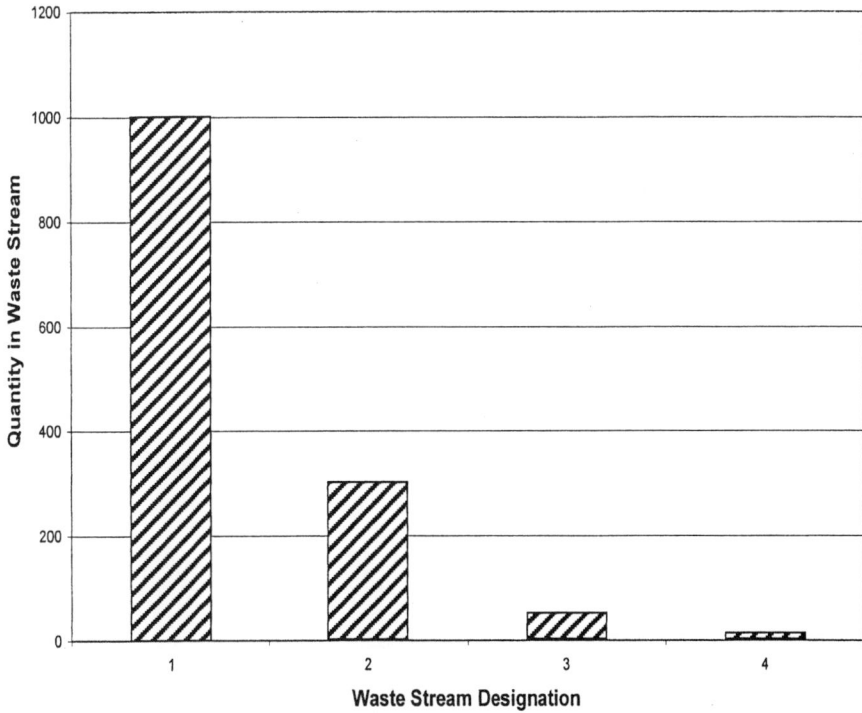

References

1) Mulholland, Kenneth L. and James A. Dyer. 1999. *Pollution Prevention: Methodology, Technologies and Practices.* New York: AIChE.

Section III

Preparation for Opportunity Identification

Introduction

Now that the waste streams have been selected, a data package has to be prepared, analyzed and transmitted to the opportunity identification team prior to the brainstorming session. Brainstorming uses the synergy of group dynamics, coupled with the ingenuity of each participant, to generate a superior set of ideas than could have been achieved by each person working alone.

Collect Data

The amount of information to be collected will depend on the complexity of the waste streams and the process that generates them. Material balances and process flow diagrams are a minimum requirement for most waste minimization opportunity identifications. The four types of desirable information to collect are described below.

Process Flowsheets and Flow Chart

Process flowsheets are the typical drawings that contain depictions of vessels such as reactors, heat exchangers, separation columns, tanks, pumps, etc (Figure III-1). The flowsheets show the major process lines and sometimes the major utility lines such as steam and brine. Experienced chemical engineers can read the flowsheets, follow the movement of products and other materials and understand the function of each major piece of equipment. However, some members of the opportunity identification team may not be trained to read and understand these flowsheets. To help those team members the process expert develops a process flow chart along with the function descriptions of the major process units depicted in the flowsheet.

A flow chart shows the major steps or functions starting with the input of the feed materials, to the product(s) formulation and separation of the product(s) from the wastes and unused feed materials (Figure III-2). For each step a function description form is completed that describes the step, lists its important operating parameters and contains other information such as maintenance frequency, unusual operating conditions, and any waste generated at that step. (See Appendix A)

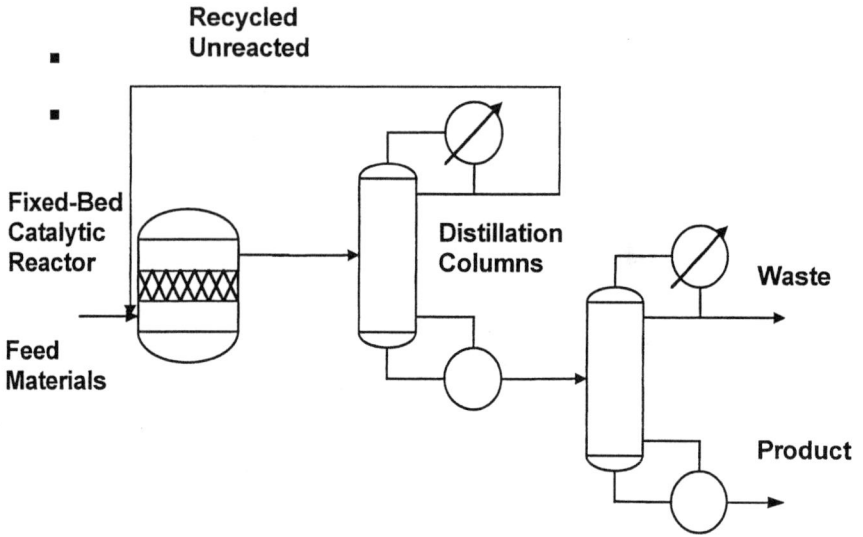

Figure III-1 Typical Chemical Flowsheet

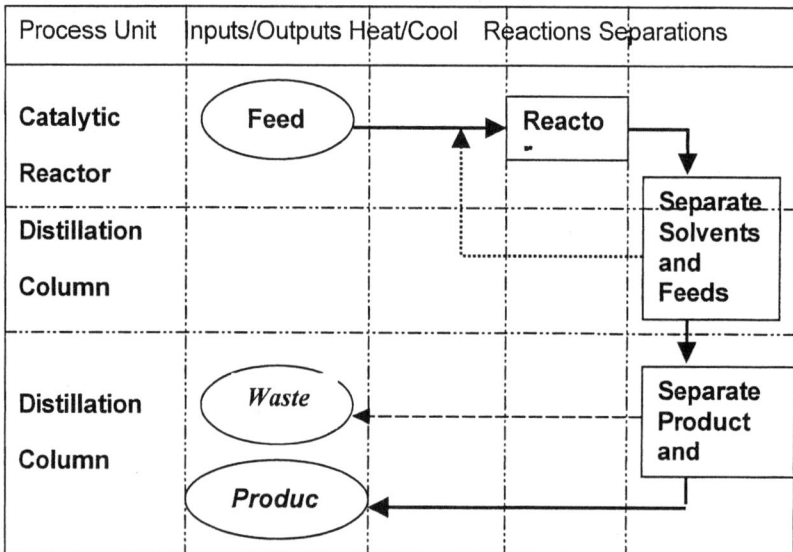

Figure III-2 Process Flow Chart

For the team member who is not familiar with the process the function description form describes each flow chart step in sufficient detail for the person to understand the function and to develop ideas to improve the step. The function description form contains:

- Flowsheet Designation

- Type of Unit (reactor, distillation column, heater, tank, and so on)

- Function of Unit

- Principal Control Parameter(s)

- Principal Poor-operation Problems

- Uptime (Average time between outages)

- Wastes going to treatment or being emitted to the environment

Heat and Mass Balances

The second necessary information to include in the data package is a listing of the materials in the process streams. The stream information in Table III-1consists of:

- Identification as shown on the Flowsheet and Flow Chart

- Phase – liquid, gas or solid

- Mass and/or mole flow of each component in the stream

- Total flow

- Temperature

- Pressure

- Enthalpy (if possible)

The mass and heat balances can be prepared using spreadsheet programs or chemical process simulators such as ASPEN PLUS™, Pro/II™, etc. A source of information is operating manuals with process descriptions (particularly valuable for batch processes).

Table III-1 Stream Mass Flows

Stream Number	A1	A3	A4	A5	A6
Phase	Vapor	Vapor	Vapor	Liquid	Liquid
Components					
Flows in KG/HR					
Benzene	0	0	0	0	0
Nitrogen N2	16761.4272	16761	16761	0	0.0998
Oxygen O2	5124.7728	5018.6	5018.6	0	0.024
Water H2O	134.2656	196.54	196.54	7938	6722.8
REACtant	219.0888	93.351	93.351	0	91.99
PRODuct	0.4536	94.213	94.213	0	94.213
IMPURity	6.804	6.804	6.804	0	1.0886
Ammonia NH3	18.144	3.6288	3.6288	0	0
Nitric Dioxide NO2	0	6.3504	6.3504	0	1.0433
Sulfur Dioxide SO2	0	17.736	17.736	0	1.3154
Carbonyl Sulfide COS	0	8.3009	8.3009	0	0.0008
Carbon Dioxide CO2	0	54.84	54.84	0	0.0113
Hydrogen Cyanide HCN	0	7.4844	7.4844	0	1.9505
Sulfuric Acid H2SO4	0	0	0	0	11.794
Ammonium Sulfate	0	0	0	0	13.608
Total	22264.956	22269	22269	7938	6939.9
Temperature C	266.5	400	352	30	54.2
Pressure kg/sqcm abs.	5.2725	2.1793	2.0387	1.7575	1.2654
Molecular Weight	28.9	28.9	28.9	18	18.4
Density kg / cu. m.	3.363864	1.1053	1.1053	1007.6	989.94

Process Chemistry

The third necessary information to include in the data package is a description of the materials in the process, chemical reactions and reactor operating conditions. This includes

- Key physical and chemical properties for the feed materials, catalysts, byproducts, and products,

- Definition of the chemical reactions (sequential or parallel), particularly noting if the desired product is an intermediate in the reaction sequence,

- Acceptable reactor operating conditions,

- How reactants and products are added to and removed from the reactor,

- Trace impurities,

- Equilibrium relationships,

- etc.

Component Information

The fourth necessary information to include in the data package is a listing of all materials added to the process, such as solvents as well as feeds, reaction intermediates and products. The minimum information for each component is (See Appendix A):

- Formula

- Molecular weight

- Density

- Normal boiling point

- Normal freezing point

Team Data Package

The data required for the brainstorming session was discussed in the "Collect Data" step above. The core team prepares a data package to distribute to all brainstorming participants at least two weeks before the session. This package should include

- Process Flowsheet

- Process Flow Chart

- Function Description Forms

- Heat and mass balances of all the major process streams

- Identification of all waste streams, including available information on flow rate and composition (Waste Stream Description Forms from Section II)

- Process chemistry

- Component information

- Supplementary information that can be included is identification of the major energy consumers in the process, and identification of any special hazards.

Do not overwhelm the group with a large volume of information. However, the package needs to contain enough information that group members, who are not part of the core team, can come up to speed quickly.

Define the Problem

This step in the preparation for opportunity identification helps the team to understand the targeted waste streams as well as the process steps that generate them. Two activities are strongly recommended, an area inspection and waste stream and process analyses.

Perform an Area Inspection

Experience has shown that a plant tour is a good way for those team members who are not familiar with the process to improve their understanding of the processes and equipment pieces that generate waste. Conduct the inspection when the process is up and running, especially the process areas of concern.

Perform Waste Stream and Process Analyses

Section I introduced two techniques—Waste Stream Analysis and Process Analysis—to help parse or divide the overall process and individual waste streams into their important parts. The goal of these two analysis techniques is to frame the problem, such that pertinent questions arise. Experience shows time and time again that when the right questions are asked, the more feasible, practical, and cost-effective solutions for waste minimization become obvious. This is the wisdom of waste minimization —that properly defining and segmenting the problem ultimately leads to the best waste minimization solutions. Analyzing the manufacturing process in this manner before and during the brainstorming session will often result in an improved process that approaches zero waste generation and emissions.

Waste Stream Analysis

The best waste minimization options cannot be implemented unless they are identified. The identified options will eliminate or minimize the waste streams volumetric flo (investment and operating costs) or eliminate / minimize the components of concern (the reason to treat the stream). To uncover the best options, each major waste stream for which a WasteStream Description Form was filled out in Section II should be analyzed with these four steps:

1) List all components in the waste stream, along with any key parameters. For instance, for a wastewater stream these could be water, organic compounds, inorganic compounds (both dissolved and suspended), pH, etc.

2) Identify the components triggering concern, *e.g.*, hazardous air pollutants (HAPs), carcinogenic compounds, wastes regulated under the Resource Conservation and Recovery Act (RCRA), etc.

- Determine the sources of these components within the process.

- Then develop process improvement options to reduce or eliminate their generation.

3) Identify the highest volume materials — often these are diluents, such as water, air, a carrier gas, or a solvent. These materials frequently control the investment and operating costs associated with end-of-pipe treatment of the waste streams and have a significant impact on the process cost of manufacture.

- Determine the sources of these high-volume materials within the process.

- Then develop process improvement options to reduce their volume or eliminate them.

4) If the components identified in Step 2 are successfully minimized or eliminated, identify the next set of components that have a large impact on investment and operating cost (or both) for end-of-pipe treatment. For example, if the aqueous waste stream was originally a hazardous waste and it was incinerated, eliminating the hazardous compound(s) may permit the stream to be sent to the wastewater treatment facility. However, this may overload the biochemical oxygen demand (BOD) capacity of the existing wastewater treatment facility, making it necessary to identify options to reduce organic load in the aqueous waste stream.

Process Analysis

The manufacturing facility in Figure I-3 represents a case where all of the materials added to or removed from the process are valuable to the business. Other than the feed materials, the input streams shown in Figure I-2 (e.g., catalysts, air, water, etc.) are required because of the designers' limited knowledge of how to manufacture the product without them. The *function* of the catalysts, solvents, water, etc. is to transform lower-value feed materials to higher-value products. To attain the process shown in Figure I-3, either the feed materials, intermediates, or products must serve the same *function* as the other input materials, or the process needs to be modified to eliminate them. To help frame the problem for an existing manufacturing facility, then, a Process Analysis should be completed.

For either a new or existing process, the following steps are taken to conduct a Process Analysis (See Appendix A for Process Constituents and Sources form):

1) List all feed materials reacting to salable products, any intermediates, and all salable products. This is "List 1."

2) List all other materials in the process, such as non-salable byproducts, solvents, water, air, nitrogen, acids, bases, and so on. This is "List 2."

3) For each compound in List 2, ask "How can I use a material from List 1 to do the same function of the compound in List 2?" or "How can I modify the process to eliminate the need for the material in List 2?"

4) For those materials in List 2 that are the result of producing non-salable products (i.e., waste byproducts), ask "How can the chemistry or process be modified to minimize or eliminate the wastes (for example, 100% reaction selectivity to a desired product)?"

When coupled with the application of fundamental engineering and chemistry practices, examining a manufacturing process by the Waste Stream Analysis and Process Analysis techniques will often result in a technology plan for driving toward a minimum waste generation process.

Show Stoppers

If you determine that incoming materials are a significant cause of waste generation in your process, STOP and consider whether to expand the scope of the waste minimization program to include an upstream process. If you are unable to determine the cause of a significant fraction of the waste, STOP and consider whether any additional studies should be performed before proceeding any further with the opportunity identification phase.

Upstream Process pH Control

A reaction/distillation process on a plant site generated a large waste stream consisting mostly of unrecovered product. Extreme pH variability in one of the incoming feed materials to the process frustrated efforts to reduce the waste stream by masking the true causes of waste generation. The only alternative available to the process engineers was to investigate changes in distillation control parameters such as temperature, pressure, and flow rate. These changes were never implemented, however, because their waste reduction potential was small.

There was, however, another alternative. The process area decided to involve the upstream process area in their waste reduction program. As a result of this cooperative approach, the upstream process area upgraded their pH control system and greatly reduced feed material variability. It was not long after this change that the downstream reaction/distillation process area identified and implemented a number of worthwhile waste reduction options in the reaction step—the main source of the waste. As a result, the downstream process area achieved a 60% reduction in waste generation resulting directly from looking at the upstream process. You may find that you need to move upstream to the supplier of a key feed material to successfully attack the cause of waste generation.

Preparation to Generate Options

When the core team gains a good understanding of the process and the source and cause of each waste stream, the team should convene a meeting to brainstorm for ideas. In order to generate <u>all</u> possible ideas for process improvement and waste reduction, the core assessment team will need to be supplemented with additional talents and diverse points of view.

In addition to generating new ideas, brainstorming sessions often prove successful in building enthusiasm for older ideas that were not pursued in the past (for whatever reason) by people close to the process and problem. Because ideas are not accepted or rejected during the brainstorming session, the environment encourages a creative and non-confrontational exchange of ideas.

The goal of the brainstorming group is to identify technological changes to (1) reduce process waste generation and energy requirements and (2) reduce the cost of manufacture. To conduct an effective technology brainstorming session, careful preparation is required by the leader, facilitator and all group members.

There are four main steps in the brainstorming process:

1) Data collection and preparation of an information package for all participants as discussed previously.

2) Selection of team members

3) Information review

4) Brainstorming session

Selection of Team Members

The composition of the brainstorming group will differ depending on what type of process is being analyzed. However, in general, the group ideally will consist of the core team (4 to 6 members) plus additional members possessing the skills required to have an effective brainstorming session. The size of the total group will typically range from a minimum of 8 to a maximum of 20 team members. Experience shows that 8-15 is the best size.

The brainstorming group should include people with the following skills or expertise:

- Business leadership and knowledge—ideally, a business representative conveys upper management support for the program and emphasizes the importance of the group coming together. In many cases, the business representative will learn a great deal about the technology and manufacturing process, including its limitations.
- Process engineering—at least one process engineer from the business is required to play the role of process expert during the brainstorming session. Normally, this is a member of the core team. This expert will walk the group through the process flow diagrams explaining the whys and hows of the existing process. His/her main duty is to be the source of process knowledge, not to participate in the idea generation portion of the brainstorming session. One or more additional process engineers from the business should participate to share new ideas as well as other ideas that had been considered in the past, but were not implemented.
- Process chemistry—ideally, someone from the business R&D group should participate to share both new and old approaches to the process chemistry. A significant cause of process waste is the reaction chemistry. This includes byproduct formation as well as the use of catalysts and solvents (as carriers) in the reaction steps.

- Environmental engineering—an environmental specialist is able to apprise the group of any regulatory impacts and to provide background on impacts to existing waste treatment operations.

- Chemical separations—most waste streams consist of a high-volume carrier (such as water, organic solvents, air, or nitrogen) containing low levels of contaminants. Techniques for reducing the volume of the carrier or the contaminant level will depend on the type of separation technology being used.

- Energy conservation—if the process is a large energy consumer, then an energy specialist can suggest ways of improving energy usage within the process.

- Engineering evaluations—an engineering evaluator will serve as the primary coordinator of the process and economic information for each of the waste reduction alternatives. The evaluator's duties are to facilitate the screening and evaluation of the ideas generated during the brainstorming session. This person may or may not be the lead process engineer from the business or the leader of the core team; however, he/she should be a member of the core team.
- Process hazards—if there are any special hazards associated with the process, then process hazards expertise will be required.

- Operations and maintenance—a lead operator and maintenance person who are familiar with the day-to-day sources of waste, how operating procedures affect the amount of waste, and which waste streams are generated during startup, shutdown and maintenance activities.

- Wildcard or outside experts—for an effective brainstorming, at least two outside experts or wildcards are needed (usually an engineer and a chemist). The duties of these experts are to

- provide a broader experience base for the brainstorming group,

- question the technology and practices of the present process,

- educate the group and core team on other ways of making the product, and

- act as a catalyst to extract the best ideas residing within the minds of the group members.

What do the outside experts or wildcards bring to the table? They help everyone in the brainstorming group to think outside of their own "box." One often finds that for a particular product all competitors have essentially the same basic approach and technology for manufacturing that product. The outside experts and wildcards can bring a different perspective on how to approach any particular task; therefore, they offer the opportunity for better and newer ways to be considered. For example, assume that you are in the business of making specialty chemicals using batch reactors, and your main source of waste is off-spec product from product changeovers. In this situation, outside expertise in batch operations from the food, pharmaceutical, or agricultural products industries could bring a different perspective on ways to improve the present operation of your process.

Information Analysis

To be effective in a brainstorming session, the participants must study the information package before coming to the session. Because 100 to 200 ideas will be generated in a short time period, an unprepared participant will contribute less than a well-prepared participant. One proven approach to aid in the participant's participation is to use the waste stream and process analysis techniques. If the participants are stimulated to ask the right questions, then the best waste minimization solutions will almost always become obvious.
To ensure that the participants properly prepare for the brainstorming session, the invitation letter that accompanies the data package has to emphasize business leadership's expectations. (Sample letter in Appendix A) Improper preparation by the participants could reduce the number of "good" ideas by 50%. Any idea not presented is a lost opportunity.

The root causes for process waste are described in Section I-6. In summary there are two general categories of root causes for process waste.

- Operational—Operational root causes arise from how the process is operating versus how it should operate.

- Fundamental— Fundamental root causes arise from the chemistry, thermodynamic and engineering limitations of the process.
Note—Basic thermodynamic principles state that any material introduced to or created in a process will escape as a waste or will not be completely destroyed by an end-of-pipe treatment device.

As part of the information packet for the opportunity identification team, the facilitator, team leader and process expert need to develop a series of question for the participants to consider. The list of questions will help trigger the participant's creative thought processes.

Focus Areas of Questions for Participants?

Operating Procedures—Are the operating practices and operating procedures the same? Have the operating parameters been recently examined to determine if they are optimum?

Process Control—Does the lack of proper analyses limit process control capability? Has the operating parameters changed since the original control configuration?

Process Parameters—Do the process parameters match any new feed conditions, product specifications or environmental requirements? Has the operating points of the equipment changed due to fouling, cleaning, upsets caused by internal damage, use of new solvents, etc.? Can a feed, intermediate or product replace water as a cooling/scrubbing medium?

Product Specifications—Can any of the wastes be modified to sell as products? Can any of the wastes undergo a reverse reaction to generate the feed materials? Can the product specifications be relaxed?

Chemistry—Can a different solvent be used? Can the operating conditions be modified to obtain 100% selectivity? Has the feed addition sequence and rate been optimized? Can one of the feed, intermediate or products replace a solvent?

Equipment/ Process Modifications—Fast reactions in an inline reactor and slow reactions in the batch process? Packed column versus trayed column? Preheat or cool the feed?

New/Unique Technology—Continuous versus batch process? Series of micro-reactors to maximize selectivity? Crystallization versus distillation versus membranes versus absorption versus adsorption as a separations technology?

Each participant is expected to bring a different perspective to all ideas that are generated. This is why it is so important to pick the right mix of people for the session. The concept is to have everyone be aware of the interdependency of any one idea on the whole, and how that idea can impact other ideas.

Typical Questions For Each Participant to Consider (Appendix A)

For the process engineer: What is the life of the present process? The present product? What is the competition doing that the group should know about?

For the chemist: What are the principal factors affecting yield, conversion, and selectivity? If the reaction is reversible, can byproducts be back-reacted to the incoming feed materials or converted to other useable products? If non-salable products are homologues of the reactants or intermediates, how can they be converted and recycled? What other catalysts are possible? If excess reactants or inerts are being used, ask why? If air, water, or a solvent are being used, ask why?

For the separations specialist: If an exit gas or water stream is being generated, what other separation techniques could be used to eliminate the stream? For trace levels of contaminants, how can the separation unit operations be improved? If large amounts of energy are required, what other separation technologies are applicable? If significant heating, followed by cooling, and then reheating takes place, what other combinations of unit operations can be used to minimize energy usage?

For the environmental specialist: What are the hazardous, carcinogenic, or toxic materials in the waste and product streams that require or could require further treatment? What are the present and future (5-10 years out) environmental laws that impact the waste from this process? What end-of-pipe technologies are appropriate?

For the engineering evaluator: For the current waste streams, what are the end-of-pipe treatment costs? What is the cost of waste generation for the current process?

For the energy specialist: What are the opportunities to save energy in the process? What are the process-to-process energy exchange opportunities? What are the corporate energy goals?

For the lead operator and maintenance representative: What operating procedures are outdated or not followed? How does poor operation affect waste generation? How can startup, shutdown, and maintenance wastes be reduced?

Supplementary Information

A complete information package should include:

1) Waste minimization purpose—This is normally developed by the business leader and core team leader.

2) Brainstorming session purpose and products—A typical statement of purpose and products is:

Brainstorming Session Purpose and Products

Purpose: **To identify ideas to reduce waste generation.**

We Want to:

- Take advantage of your perspective and expertise.

- Identify all ideas that reduce the amount of waste generated.

- Use the synergy of a brainstorming session to identify cost-effective ideas.

- Address any barriers and your concerns.

Benefits: **The business can reduce emissions and waste, and improve the operation of the process with maximum return to the stockholders.**

Products:

- **Identification of the changes (technological and operational) required to improve operation of the process.**

- **Develop a prioritized list of opportunities and recommendations to be considered to reduce waste generation.**

- **Path forward.**

3) Brainstorming session agenda—This is developed by the facilitator and core team leader. A typical agenda is

Typical Brainstorming Session Agenda

Afternoon Day 1

Introductions

Review Waste Minimization Methodology and Brainstorming

Review Agenda, Purpose, Products, and Ground Rules

Business and Environmental Drivers

Process Overview and Sources of Waste

Plan Plant Tour (Could also be done in the morning)

Day 2

Brief Process Overview

Idea Generation

Develop Ranking Criteria

Rank Ideas

Day 3

Introduction to Idea Development

Development of Best Ideas

To maximize the contribution of the group, each individual group member needs to follow three ground rules—participate, be concise, and be additive.

First, each individual invited to the session is expected to participate—silence is unacceptable. All participants need to understand that their ideas, no matter how off-the-wall they may sound at first, will not be judged during the brainstorming portion of the meeting. The judgment and critique will be reserved for the screening portion of the session. The concept behind brainstorming is that one idea should lead to a new idea or build on the previous one. If a person does not speak up, then that individual's ingenuity is not being fully exercised.

Second, participants need to be concise. Ideas must be conveyed clearly and completely. Answer any questions on the meaning of your idea, but do not engineer the idea. There just is not enough time during the brainstorming portion of the meeting to engineer every idea. In addition, it will restrict the flow of new ideas.

Third, participants should be additive and avoid critiquing other people's ideas. Sometimes, an idea that was tried in the past and failed for either technological or political reasons will work in the current climate. Also keep in mind that all ideas will be reexamined at a later date. The goal of the brainstorming session is to get all possible ideas on the table, so that the best idea can be evaluated and chosen during the screening and evaluation stages of the assessment phase.

Outside Experts

Outside experts are invited to participate in the program to provide other viewpoints on how to improve the process. Thus, the experts have a special obligation to be prepared and additive. To be effective the outside expert should have:

- Reviewed the information in sufficient detail to understand the process,

- Developed a minimum of two improvement opportunities that address the basic root causes involving the chemistry, thermodynamics and engineering of the process,

- Communicated with the core team for any questions about the process.

An unprepared participant retards the brainstorming process by requiring the process expert to educate that person during the brainstorming session.

Process Engineers and Chemists

Process engineers and chemists who are participating in the brainstorming session have a special duty to be open-minded. If an idea is suggested that was tried in the past, the engineers or chemists cannot make any comment on the value of the idea. The fact that the idea was tried in the past and did not work is not relevant to the brainstorming. The idea could be the start for another participant to develop an even stronger idea. Also, the fact that the idea did not work in the past does not mean that the idea would not work now. Thus, the process engineer and chemist must not say "We tried that idea in the past, and it did not work."

4) **Participant responsibilities (Appendix A).**

5) List of question to consider while preparing for the brainstorming session

6) Finally include any waste minimization ideas that were previously considered.

7) Copies of chapters 7 to 18 of the book "Pollution Prevention: Methodology, Technologies and Practices" if pertinent. A review of these chapters will determine if any of the chapters contain technologies and practices that will help the participants develop improvement ideas. The titles are:

- Chapter 7: Pollution Prevention in Batch Operations

- Chapter 8: Equipment and Parts Cleaning

- Chapter 9: High-Value Waste

- Chapter 10: Reactor Design and Operation

- Chapter 11: Use of Water as a Solvent and Heat Transfer Fluid

- Chapter 12: Organic solvents

- Chapter 13: pH control as a Pollution-Prevention Tool

- Chapter 14: Pollution Prevention in Vacuum Processes

- Chapter 15: Ventilation of Manufacturing Areas

- Chapter 16: Volatile Organic-Liquid Storage

- Chapter 17 Separation Technology Selection

- Chapter 18: Equipment Leaks: Regulations, Impacts, and Strategies

Summary

The information package should be sent to the participants one to two weeks before the brainstorming session. The participants will have the responsibility to review the information and come prepared to participate and offer improvement ideas.

1) Cover letter with an invitation to participate and containing the when and where for the 2 to 3 day meeting. Letter needs to be sent by the core team leader.

2) Introduction containing—

- Brief description of the process including the overall chemistry,

- Process waste calculation as described in Section II,

- Minimum waste calculation

- Lost production value calculation

3) Waste Minimization Program Purpose that is developed by the business and core team leaders and should include—

- Desired goals of the cleaner production program that will be used as criteria to rank the ideas that are generated during the brainstorming session,

- Introduction to brainstorming.

4) Brainstorming Session Purpose and Products

- Developed by the core team leader and facilitator.

5) Brainstorming Session Agenda

- Developed by the core team leader and facilitator.

6) Participants Responsibilities

7) Problem Definition

- List of initial questions

- Waste stream analyses

- Process analysis

- Typical questions for each participant to consider

8) Team Data Package

- Design drawing or sketch of the overall process.

- Process Flow Chart

- Function description on the purpose of each major piece of equipment.
- Stream Heat and Mass Balances for streams on the sketch

- **Process Chemistry**

- **Component Information**

9) If pertinent, wastewater treatment facilities and discharges, vapor vent treatment devices and solid waste treatment.

- Provided by the team leader, process engineer and environmental professional.

10) Identification of the compounds of concern and timing for reduction.

- Provided by the team leader and environmental professional.

11) Any waste minimization ideas that were previously considered

12) Copies of any appropriate chapters from the book "Pollution Prevention: Methodology, Technologies and Practices."

Section IV

Opportunity Identification

Introduction

The heart of the waste minimization methodology is the opportunity identification brainstorming session. The process improvement ideas that are generated during this typically 4 to 8 hour session range from simple procedural changes to those requiring some level of engineering effort or, in some cases, research and development. In effect, this list of process improvement ideas describes the technology needs for the business over the next 5 to 10 years. It also identifies the technological limitations of the existing process.

Set Goals

This task helps the team to analyze the drivers for waste minimization and to develop the criteria needed down the road to screen the options generated during the brainstorming session. Some examples of goals from which screening criteria can be developed include:

- Technical feasibility; that is, will the option work?

- Little or no capital investment is required.

- Minimal plant downtime is needed to implement the option, that is, fast turnaround time.

- Energy usage is lowered.

- Plant operability is improved.

- Waste generation and emissions are reduced.

- Operating costs are lowered.

- Regulatory objectives are met or exceeded.

The Brainstorming Session

Brainstorming uses the synergy of group dynamics, coupled with the ingenuity of each participant, to generate a superior set of ideas than could have been achieved by each person working alone. To be effective, therefore, a brainstorming session must be well-structured and "friendly" at the same time. According to the playwright Eugene Ionesco, "it is not the answer that illuminates, it is the question." With the proper atmosphere, questioning will generate a positive rather than a defensive response.

Logistics

In the information package sent to the participants, a cover letter should be included describing the drivers, purpose, expected products, participant responsibilities, and agenda.

Drivers

Normally, both external and internal business drivers exist for doing waste minimization. The premier external driver is usually regulatory, which means that there will likely be a significant incentive to avoid investment in new or upgraded end-of-pipe treatment. Other external drivers include public opinion, desired greenness of the product or process, proximity of the facility to the residential community, and market share.

Internal business drivers can include corporate or business goals and vision, realization of the cost of waste generation to the business, and the need to increase capacity in existing facilities while avoiding permitting barriers to growth.

Purpose and Products

The purpose and products statement outlines the "why, how and what" of the meeting, together with the expected products of the brainstorming session itself.

Participant Responsibilities

A short discussion of participant responsibilities should be included in the information package. These responsibilities should also be reviewed at the beginning of the brainstorming meeting. The key responsibilities are to participate, be concise, and be additive.

Agenda

A typical brainstorming session will last 2-1/2 days beginning around noon the first day. During the afternoon of the first day, the purpose, products, drivers, and ground rules are reviewed, business leadership provides a business overview, and the lead process engineer will introduce the manufacturing process using process flow diagrams, flow charts and an optional plant tour. The process overview should emphasize unusual process conditions that affect waste generation and highlight each waste stream of interest, including how it is generated. The group must finish the first day with a good understanding of how the process currently operates. The goal is to lay the groundwork for the generation of ideas the next morning.

The second day is a full day. It begins bright and early with a quick review of the process, followed immediately with the brainstorming of ideas. This will usually last 4-6 hours. After all ideas have been generated and recorded, the group will develop or review (preferred) two or three screening criteria which they will use to perform a first-cut ranking of the ideas. Ideally, the idea generation period and ranking period should be roughly the same length of time; however, experience has shown that this can usually be accomplished in 1 to 2 hours. Quite frankly, we find that because everyone is so exhausted from the idea generation stage, they are anxious to go home when it comes time to rank the options. For this reason, it is important to prepare the screening criteria ahead of time and to simply seek the group's buy-in before the ranking.

The third day is also a full day. Each of the highest ranked ideas are assigned a champion. The champion will then do an initial high level engineering evaluation of the idea during the third day. By doing the evaluation with the other participants present accomplishes:

1) Interaction with everyone on the details of the idea including any clarifications or modifications.

2) Education of everyone on the more radical, but potentially, more valuable ideas. The person who advocated a more radical approach will have a chance to educate and persuade other members of the team on the value of the idea.

Ground Rules

The ground rules are designed to help control the atmosphere in the room during the session. The ground rules are an extension of the participant responsibilities and will be agreed upon by the participants at the beginning of the meeting. They should be posted in the meeting room and used to facilitate an effective and productive session.

Typical Ground Rules for a Brainstorming Session (Appendix A)

Participate — All ideas are good ideas.

Stay Focused — Keep the business needs and purpose of the brainstorming session in mind.

Build on Ideas — Use other people's ideas to create synergy.

Be Polite — Listen to understand; the person talking has the floor.

Be Positive — Work to sharpen the ideas being generated.

Recording of Ideas

During the generation of ideas, someone will need to record or scribe each idea in 1-2 lines on a flip chart.

Brainstorming Responsibilities

The goal of the brainstorming group is to identify technological changes to

- Reduce process waste generation and energy requirements, and

- Reduce the cost of manufacture.

To conduct an effective technology brainstorming session, careful preparation is required by all.

In addition to generating new ideas, brainstorming sessions often prove successful in building enthusiasm for older ideas that were not pursued in the past (for whatever reason) by people close to the process and problem. Because ideas are not accepted or rejected during the brainstorming session, the environment encourages a creative and non-confrontational exchange of ideas.

Participants

During the brainstorming session, anywhere from 100 to 200 ideas will be generated over a 2 to 8 hour period. This is a lot of ideas! To maximize the contribution of the group, each individual group member needs to follow three ground rules—participate, be concise, and be additive. The concept is to have everyone be aware of the interdependency of any one idea on the whole, and how that idea can impact other ideas.

Outside Experts

Outside experts are invited to participate in the program to provide other viewpoints on how to improve the process. Thus, the experts have a special obligation to be prepared and additive. To be effective the outside expert should have:

- Reviewed the information in sufficient detail to understand the process,

- Developed a minimum of two improvement opportunities that address the basic root causes involving the chemistry, thermodynamics and engineering of the process,

- Communicated with the core team for any questions about the process.

An unprepared participant retards the brainstorming process by requiring the process expert to educate that person during the brainstorming session.

Process Engineers and Chemists

Process engineers and chemists who are participating in the brainstorming session have a special duty to be open-minded. If an idea is suggested that was tried in the past, the engineers or chemists cannot make any comment on the value of the idea. The fact that the idea was tried in the past and did not work is not relevant to the brainstorming. The idea could be the start for another participant to develop an even stronger idea. Also, the fact that the idea did not work in the past does not mean that the idea would not work now. **Thus, the process engineer and chemist must not say "We tried that idea in the past, and it did not work."**

Facilitator

The facilitator is the person who will lead the meeting and whose tasks are

1) Adhere to the agenda,

2) Be sure the ground rules are followed,

3) Keep the session open and friendly, and

4) Support the scribes and process expert(s).

After the idea generation stage is complete, then the ideas have to be ranked. All ideas are saved. The purpose of the ranking is to help the business focus on those ideas that will meet the purpose of this review and will have the highest return to the business.

As each idea is examined, the facilitator:

5) Asks for comment from the person who proposed the idea. All of the ideas will be included in the report. Thus if this idea is not used at this time, but if someone reviews the report a year form now to see what other ideas should be considered, then these notes become important to help identify the idea and its value. The scribe writes under the idea.

■ Any comments on the idea.

■ What is the value of the idea, that is, waste reduction, recovered product, lower raw material use, lower energy use, increased uptime, lower operating cost, etc.

6) Asks for discussion by anyone else.

7) Determines if there are other ideas that have been generated that are part of the idea under discussion.

8) Asks for a ranking of **High / Medium / Low**.

"How - Why"   Questions?

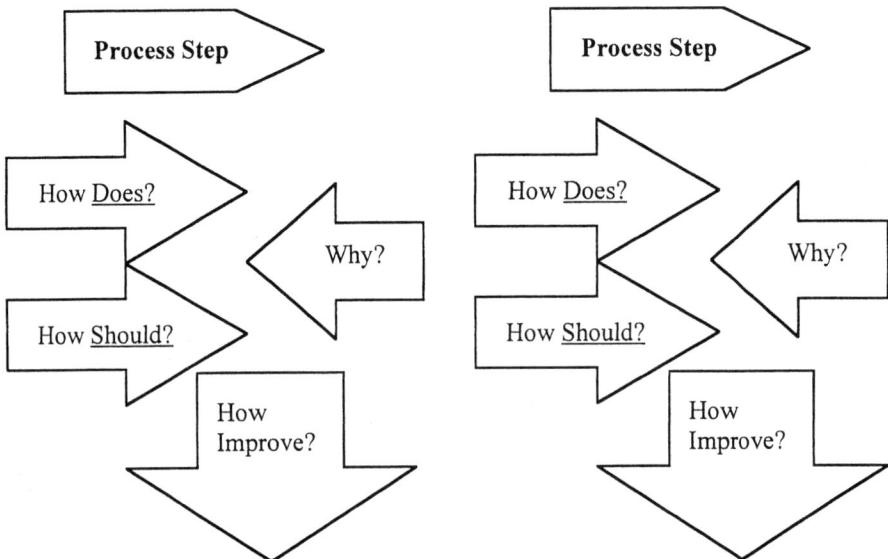

For groups with little experience in brainstorming, the facilitator may want to introduce the group to the concept of open-ended thinking and the synergy between group dynamics and idea generation. For example, *how* does versus how <u>should</u> that particular process step function? *Why* does the process step not function as it <u>should</u>? *How* can the process step or flow streams be modified to improve its operation and minimize waste generation?

The asking of "why" three to five times is another technique that leads to the root cause of a problem and thus identification of the best opportunities for improvement. The early one to two "whys" deal with the surface and operational causes of the waste generation. As more "whys" are posed then the fundamental thermodynamic, engineering and chemistry root causes are discovered.

Process Expert

The process expert(s) provide the process knowledge which enables the group to

- Understand each aspect of the process,

- Help interpret ideas as they are being generated, and

- Make sure that no part of the process is overlooked.

The process expert should have prepared transparencies of the various parts of the process diagram to lead the participants through the process.

Scribes

The scribes are the unsung heroes of the meeting. Their duties are to

1) Record each idea in a concise, yet complete, way,

2) Check with the generator of the idea to make certain that the text description is accurate,

3) Avoid editing the idea,

4) Number each idea and page and record the initials of the person who provided the idea, and

5) Group the ideas (if possible) by process area, equipment piece or flowchart function area.

Example:

Byproduct Recovery

1) Idea to change or do something   ANC (initials of person)

2) Another idea to change or do something   EKL (initials of person)

3) Another idea to change or do something   GBA (initials of person)

Reactor

4) Another idea to change or do something   WEK (initials of person)

5) Etc.

During the ranking of the ideas the scribe will

A) Record any additional comments generated during the discussion.

B) Record **H** for high, **M** for medium and **L** for low.

C) Note the number of any other ideas that are similar or are parts of the
same idea.

Idea-Recorder

If the business has enough resources, then an idea-recorder can be used
to enhance the recording of the ideas. The idea-recorder would input the
ideas into a PC as they are being generated. The recorder would input as
much of the information as possible; however, that person would not
impede the idea generation process to complete the notes. During the
screening portion of the brainstorming session the recorder can complete
the description of the ideas.

Brainstorming Room Setup and Supplies

Setup

- U shaped tables or a single large table. The participants should be
  able to see each other.

- Two chart pads with pens.

- Wall space to attach sheets of ideas.

- Transparency projector.

Supplies

- Refreshments for the breaks.

- Tape or pins to attach sheets from the chart pads to the walls.

- Tent cards for everyone to write their name. The outside experts and facilitator probably will not know everyone.

- Clear transparencies plus transparency pens for the participants to sketch their ideas.

- Chart pad sheet with the brainstorming rules printed on them and posted on the wall.

Idea Generation

The facilitator and process expert can use one of two techniques to help focus the idea generation.

1) One approach is to divide the process flow diagram into smaller blocks during the brainstorming. This allows the participants to focus their idea generation on a particular area or unit operation of the process.

2) A second approach is to focus on a particular waste stream, and then work back through the process tracing the generation of that waste.

The goal is to focus on those process steps or chemistries that lead to the generation of unwanted materials in that stream. For example, you may trace the nitrogen in a contaminated air stream from a scrubber back to the partial oxidation reactor where air was used as the source of oxygen.

A technique that works very effectively is to start with the waste stream and work back up through the process. As each block or function step is encountered questions are posed to initialize the creativity process and idea generation. The goal is to focus on those process steps or chemistries that lead to the generation of unwanted materials in those streams. For example, you may trace the nitrogen in a contaminated air stream from a scrubber back to the partial oxidation reactor where air was used as the source of oxygen.

The two reasons why starting from the back end of the process and working your way through the process generates new ideas are:

1) Most processes have never been examined in this manner.

2) In most processes the reactor system and major separations units have been studied extensively. The waste treatment facilities and its upstream separation's units are less important to the process experts, and thus there is much less resistance when new improvement ideas are identified using the methods described in this manual. As the group moves up closer to the important parts of the manufacturing process, the process expert begins to feel more and more threatened. However, the more that person understands that his/her ability is not being questioned, then the greater chance of an acceptance of new ways to improve the performance of those parts of the process.

To make sure that all possible causes are considered the facilitator or process expert will remind the participants of the categories where the possible causes and solutions lie—the same categories that would be used to list the ideas in an Ishikawa construction (cause and effect or wishbone diagram).

- Operating procedures

- Process control

- Process parameters

- Product specifications

- Chemistry

- Equipment / process modifications

- New / unique technology

Screening the Options

Once the idea generation phase is complete, the ideas need to be reexamined and ranked based on a set of criteria. The goal in this step is to perform a first-cut screening of the options. Ideally, the number of ideas that are considered to be worthy of further evaluation will be cut in half. The first-cut screening is completed during the brainstorming meeting because not all the participants are members of the core assessment team. The benefits of this are that

1) the ideas are fresh in everyone's mind,

2) those participants who generated the ideas are in the room and can add clarifications, if needed,

1) the discussions that ensued during the idea generation permit a "gut feeling" of how each idea should be ranked ranks relative to the others, and

2) you may never assemble this much expertise in one room at the same time again.

Remember that no idea is ever permanently discarded, and the core team has the responsibility to record and consider all ideas. However, the first-cut screening helps the brainstorming team to focus on the most attractive ideas to be evaluated the next day.

Normally, 50 to 200 ideas will result from the brainstorming. This is far too many for limited business resources to digest and evaluate. During the first-cut screening, a number of ideas will be combined, duplicates eliminated, and a fair number screened out. To accomplish this task, the group needs to agree on the criteria; typically no more than three criteria are used. The core team will develop the criteria from the "Set Goals" discussion. This was discussed in the front of this section.

Avoid using more than 3 criteria. Because everyone will be tired by the end of the second day, they will be unable to process more than 3 criteria anyway. Keep it simple—use **high**, **medium**, or **low** against the criteria.

Screening Technique

Starting with first idea the facilitator or process expert asks the person who generated the idea to:

1) Provide any needed explanations to complete the idea, and

2) Discuss the value of the idea.

During the conversation the scribe further expands the idea with key words under the idea to help future documentation of the idea. The group ranks the idea High, Medium, or Low.

Section V

Opportunity Evaluation and Final Report

Introduction

The failure of many waste minimization programs can be traced to the inability of the engineers and scientists involved in the programs to convince business leadership to change the manufacturing process. Often, this reluctance to change is not because the recommended process improvements were not technically sound, but because of the team's failure to speak the language of the business person—that is, dollars and cents. The role of economics in waste minimization is very important; even as important as the ability to identify technology changes to the process.

Economics plays a key role in evaluating and ranking the screened options from the brainstorming session. In other words, what is the economic viability of the best ideas? How will these changes make the business more profitable? After all, this is what business leadership is most concerned about—making money. In this section, we will show you how to use engineering evaluations techniques, which includes shortcut net present value calculations, to determine the economic value of the technically feasible options to the business. These techniques will help you to speak the language of the business people, so that the best ideas are seen by the business people as business opportunities.

Evaluate "Best" Opportunities

The next step will be to assign a champion to each of the best ideas. The champion will then do an initial high level engineering evaluation of the idea during the third day of the brainstorming session. By doing the evaluation with the other participants present accomplishes

1) Interaction with everyone on the details of the idea including any clarifications or modifications.

2) Education of everyone on the more radical, but potentially, most valuable ideas. The person who advocated a more radical approach

will have a chance to educate and persuade other members of the team on the value of the idea.

Economic Criteria for Technology Comparisons

Technology evaluation encompasses not only the technical feasibility of a particular technology, but also the economics of its implementation. While there are many measures of economic merit, the two measures—Net Present Value (NPV) and Investment—provide the business person with a complete set of information to make an informed, economic decision. Investment refers to how much money must be spent initially to design and build the new facilities. Net present value is the total value or cost (if negative) of the alternative to the business. It includes all contributions to the cash flow of the alternative—new investment, working capital, cash operating costs, income taxes, and revenues.

Net Present Value

NPV is the after-tax worth in today's dollars of all the future cash that an alternative will either consume or generate. NPV includes the effects of the four primary economic elements: investment, cash costs, revenues, and taxes. It is the most popular single measure of the economic merit of an alternative.

Investment

Investment includes all the monies required for plant, property, and equipment.

Evaluate the Alternatives

To choose the best alternative there is a need to estimate new capital investment, to determine the change in manufacturing costs and revenues, to calculate the net present values, to list all other "non-economic" considerations and, finally, to choose the best alternative. Figure V-1 defines the important components that should be included as new permanent investment. Figure V-2 describes the information required to estimate changes in operating costs and net present value.

Figure V-1 Factors Included in Permanent Investment

Permanent Investment

Permanent investment is often the single most important element in an economic evaluation. Permanent investment is defined as capital expenditures for plant, property, and equipment. It is the capital necessary to provide the installed equipment and all the auxiliaries that are needed for complete process operation. We must consider all the physical facilities need to manufacture and to sell our products in the marketplace, even though some of these facilities may be remote from the process under study.

Total investment is working capital plus permanent investment. Permanent investment is the sum of direct manufacturing facilities and supporting investment.

Direct manufacturing facilities include the capital expenditures for equipment and buildings directly involved in product manufacture.

Supporting investment. Manufacturing facilities cannot operate without substantial supporting permanent facilities. The supporting investments are
- Power facilities are the centralized facilities for generating or distributing electricity, steam, air, water, brine, etc. They include both the buildings and equipment necessary to provide this function.
- General facilities are the common facilities of a general nature at a plant site. They include such items as sewers, site improvements, fences, parking areas, railroads, road, walkways, and alarm, guard, and communications systems.
- Service facilities are common facilities that provide specific services necessary to the function of a plant site. Included in the category are administration buildings, cafeterias, shops, stores, sewage treatment facilities, mobile equipment, change houses, etc.
- Technical facilities includes buildings and equipment used for research and development activities or for routine product testing and control.
- Land investment is the capital required for site real estate. Because land investment is small and common to all alternatives under consideration, this category is often neglected in economic estimates of permanent investment.
- Precious materials includes those items subject to governmental regulation and/or special accounting and often are precious metals, e.g., platinum and its alloys, used to catalyze a chemical reaction.

Power, General, and Service (PG&S) facilities are usually a significant portion of the permanent investment. PG&S ranges from 15% to 40% of direct manufacturing investment for chemical processes and from 20% to 35% for polymer processes.

Figure V-2 Factors Required to Estimate An Alternative Cash Flow

---

Guidelines
Operating Costs and Economics
Description of Information Required for Estimating

Material Flows
The mass flowrate of raw materials required. Include replacement of adsorption resins and catalysts at the end of their life. Delivered prices.

Utility Flows
Utility flowrates and their incremental costs.

Miscellaneous Costs

Basic Data Development Requirements
The costs to test, demonstrate, and develop basic data for the final design and operation of the alternative. Include the dollars spent for testing, the cost of building a test facility, if required, and the technical/research people required.

Operator Requirements
The amount of operator attention required to operate the alternative. Measure in hours per 24 hour day of operation. Wage and benefit rates.

On-going Technical Support
The amount of technical/research required to support the on-going alternative operation. Measured as a percentage of a full-time person.

Other Miscellaneous Costs/"Non-Economic" Considerations
Anything such as extraordinary maintenance requirements, etc. Also, list any key technical uncertainties discovered or key assumptions made during the flowsheeting and scoping of the alternative. List any "non-economic" advantages or disadvantages of the alternative.

Income Taxes Are Important

Since combined U.S. and state income taxes are about 38% of pretax earnings, a savings of $1 nets only 62¢ in cash. Investment dollars are always after-tax.

The Past Is Irrelevant

We often confront projects on which a lot of money has already been spent. We hear such comments as, "It's time to stop pouring money down a rat hole," or "After spending so much, it's a shame to stop when with a little more we might reach our objective." Both these sentiments are wrong! The money that has been spent is gone, and "nor all your piety nor wit" can bring it back. We are concerned only with money yet to be spent. This is the money over which we still have control. Do not let emotions get in the way of practicing this principle. Exclude the past from all your evaluations.

Timing is important

Money has a time value based on its ability to earn a return, if invested. You must consider the timing of cash flows. Also, any money already spent or committed should not be included in the economics, because it cannot be "unspent."

Economic merit = business wealth

Economics should be on a total business, out-of-pocket cash basis, not on a process or plant basis. Be careful with "cost sheet" costs and allocated costs because they may not be equivalent to true company cash costs. Use a sound measure of economic merit such as net present value; never use just investment or annual costs. When optimizing the economics of an integrated alternative, optimize the entire alternative, not just each of its individual parts. Do not confuse economic merit with accounting — accounting concerns itself with how the company pie is divided among processes or plants; economics is concerned with the total size of the pie.

The future is uncertain

Factors affecting the relative economics can change over time. Consider how tightening of regulations or changes in production forecasts or specific utility costs may affect the relative rating of the different alternatives. At some point, a decision and risk analysis may be justified.

Money is not everything

There are many other important considerations that are not stated in direct economic terms, such as safety, company environmental goals, community relations, operability, technological risk, availability of people and capital, adaptability to tightening regulations, etc. They must be considered along with the differences in the economic price tag of all alternatives.

Shortcut NPV Method

One criticism of detailed cash flow and net present value (NPV) calculations is that they require a great deal of time and effort. Often, the amount of uncertainty in the cash flows does not warrant a detailed analysis, but an approximate NPV is still desired. The shortcut, or "back-of-the-envelope" method presented in this section provides a quick, approximate NPV which can be used for preliminary studies.

Data Requirements

First, in order to calculate NPV with this method, one must estimate four types of cash flows. Not all alternatives will have all four cash flows, but they are included here for completeness. Then, one selects the appropriate cost of capital to use for the discount rate and the appropriate life of the alternative. Factors are then looked up in a table, and plugged into the NPV shortcut formula shown below. The method handles details of inflation, income taxes, and depreciation automatically.

The first type of cash flow which must be estimated is new permanent investment. This is for the equipment and other facilities required by the alternative. It is expressed as project-level investment in today's dollars.

The second type of cash flow to be estimated is new annual revenues minus cash costs. Cash costs are all those incremental costs for which a business writes a check. Revenues are incremental sales dollars, either for products or co-products. Many alternatives do not have any incremental revenues. This cash flow is for a full year of operation after startup of the facilities. It is expressed in today's dollars; that is, before inflation, and before considering income taxes. Depending on the magnitude and sign of the costs and revenues, this cash flow can be either positive or negative. Be careful to keep signs straight. A cost savings is a <u>positive</u> value. Maintenance, property taxes, and miscellaneous other costs are automatically included in the calculations, based on the new permanent investment. Therefore, the primary types of costs that have to be estimated are raw materials, energy (utilities), and manpower.

The third type of cash flow accounts for a change in working capital. Most minimization alternatives do not significantly change product and raw material inventories, it is included here for completeness. The most common case where a significant change occurs is when a solvent is eliminated from the process. Seasonal businesses, such as agricultural products, may need to consider significant changes in working capital among alternatives. It is expressed as a change in working capital for a full year of operation and in today's dollars.

The last type of cash flow is for a one-time cash cost. This may be required for large feasibility studies. It is expressed in today's dollars. No recurring costs, such as maintenance, are factored from this one-time cost, so exercise caution when using this term.

The life of the alternative and the cost of capital must then be decided upon. These have a large impact on the NPV. The appropriate life will frequently be determined by the expected life for the waste minimization facility, until it is no longer useful, either technically or due to regulatory pressures. It may also depend on the expected life of the rest of the plant or business, depending on whether the product is a core chemical process, or a high-tech product that will become obsolete in a few years and the plant shut down.

The appropriate cost of capital should be determined by the business. If the business is aggressively pursuing waste minimization alternatives, it may choose to use a relatively low hurdle rate, even as low as the normal cost of capital, 12%. On the other hand, if the business' capital expenditures are constrained, even waste minimization alternatives may have to meet a relatively high hurdle rate, such as 25% to 30%. If a discount rate other than the cost of capital is used, it must be stated explicitly such as NPV_{25} for a 25% discount rate.

Calculating NPV

Having all the cash flows as well as the facility life and cost of capital, the NPV can be calculated. Each nonzero cash flow is multiplied by a coefficient which is tabulated as a function of the life and cost of capital. These terms are then added together to get the NPV, as shown below:

NPV = (a*Investment) - (b*(Savings + Revenue)) - (c*Δ Working Capital) + (d * One Time Cost)

Tables of the four coefficients (a, b, c, and d) as a function of facility life and cost of capital are given in Tables V-1 through V-4. The following abbreviations are also used:

I = Investment

CS = Cost Savings (cost is negative)

R = Revenue

WC = Working Capital Change (Δ) (decrease is positive, increase is negative)

OTC = One Time Cost

For a typical small project (less than $5 million investment), assuming 12% cost of capital and a 10-year life, the shortcut NPV equation reduces to

NPV_{12} = -0.91 * I - (-4.1) * (CS + R) - (-0.63) * WC + (-0.60) * OTC

This shows that a dollar of additional investment has an NPV of -$0.91, while a dollar of <u>annual</u> cost ($1/yr) has an NPV of -$4.10. The ratio of these two coefficients (-0.91/-4.1) shows that if a 12% cost of capital is acceptable, a dollar of investment can be justified by about $0.22 of annual cost savings or revenues. This is handy number to remember.

For the case of 25% cost of capital and 10-year life, the equation reduces to

$$NPV_{25} = -0.80 * I - (-2.3) * (CS + R) - (-0.67) * WC + (-0.52) * OTC$$

If 25% is the appropriate cost of capital to use, a dollar of investment can be justified by about $0.35 (-0.80/-2.3) of annual cost savings or revenues. Thus, it is substantially more difficult to justify investment based on cost savings at a 25% cost of capital than it is at a 12% cost of capital.

Basis of the Shortcut Method

The shortcut NPV method is based on a full, detailed cash flow calculation, using a particular set of assumptions. Under these assumptions, the shortcut NPV method gives exactly the same result as a full NPV calculation. It is mathematically equivalent and rigorous. However, if the assumptions are not correct, the shortcut method is only an approximation. The equation parameters are further refined by the type of project—large, midsize, and small improvement. The assumptions and other characteristics of each project type are shown in Table V-4. Each type of project has its own table of coefficients for use in the shortcut NPV equation. These coefficients are listed in Tables V-1 to V-3. To estimate the NPV of an alternative using the shortcut method, you need only to

- Identify project type,

- Choose the appropriate discount rate, and

- Select the desired facility life (labeled as "Lives" in the tables).

Table V-1 Shortcut NPV Equation Coefficients
Large Projects (>±$20 million)

Coefficient "a" (NPV of $1 of Investment)

Discount Rates

Lives	10%	12%	15%	20%	25%	30%	35%	40%
5	-0.85	-0.82	-0.77	-0.69	-0.63	-0.57	-0.52	-0.48
10	-1.0	-0.94	-0.86	-0.75	-0.67	-0.60	-0.54	-0.49
15	-1.1	-1.0	-0.92	-0.79	-0.69	-0.61	-0.55	-0.49
20	-1.2	-1.1	-0.96	-0.80	-0.69	-0.61	-0.55	-0.49
25	-1.3	-1.1	-0.98	-0.81	-0.70	-0.61	-0.55	-0.49
30	-1.3	-1.2	-1.0	-0.81	-0.70	-0.61	-0.55	-0.49

Coefficient "b" (NPV of $1 of Cost Minus Revenue)

Discount Rates

Lives	10%	12%	15%	20%	25%	30%	35%	40%
5	-2.0	-1.8	-1.6	-1.2	-0.97	-0.78	-0.63	-0.51
10	-3.8	-3.3	-2.7	-1.9	-1.4	-1.1	-0.84	-0.66
15	-5.1	-4.3	-3.3	-2.3	-1.6	-1.2	-0.90	-0.69
20	-6.1	-5.0	-3.8	-2.5	-1.7	-1.2	-0.92	-0.70
25	-6.9	-5.5	-4.0	-2.5	-1.7	-1.2	-0.92	-0.70
30	-7.5	-5.8	-4.2	-2.6	-1.7	-1.2	-0.92	-0.70

Coefficient "c" (NPV of $1 of Working Capital Change)

Discount Rates

Lives	10%	12%	15%	20%	25%	30%	35%	40%
5	-0.23	-0.25	-0.26	-0.26	-0.25	-0.23	-0.21	-0.19
10	-0.49	-0.50	-0.50	-0.46	-0.41	-0.36	-0.31	-0.27
15	0.69	-0.68	-0.65	-0.56	-0.48	-0.40	-0.34	-0.29
20	-0.84	-0.80	-0.73	-0.61	-0.51	-0.42	-0.35	-0.29
25	-0.95	-0.89	-0.79	-0.64	-0.52	-0.42	-0.35	-0.29
30	-1.0	-0.95	-0.82	-0.65	-0.52	-0.42	-0.35	-0.29

Coefficient "d" (NPV of $1 of One-time Cost)

Discount Rates

Lives	10%	12%	15%	20%	25%	30%	35%	40%
For Any Life	-0.56	-0.54	-0.52	-0.47	-0.44	-0.40	-0.37	-0.35

Table V-2 Shortcut NPV Equation Coefficients
(>±$5 million) Midsize Projects (<±$20 million)

Coefficient "a" (NPV of $1 of Investment)

	Discount Rates							
Lives	10%	12%	15%	20%	25%	30%	35%	40%
5	-0.85	-0.82	-0.78	-0.73	-0.67	-0.63	-0.58	-0.55
10	-0.99	-0.94	-0.88	-0.79	-0.72	-0.66	-0.61	-0.56
15	-1.1	-1.0	-0.93	-0.82	-0.73	-0.67	-0.61	-0.57
20	-1.2	-1.1	-0.97	-0.84	-0.74	-0.67	-0.61	-0.57
25	-1.2	-1.1	-0.99	-0.84	-0.75	-0.67	-0.61	-0.57
30	-1.3	-1.2	-1.0	-0.85	-0.75	-0.67	-0.61	-0.57

Coefficient "b" (NPV of $1 of Cost Minus Revenue)

	Discount Rates							
Lives	10%	12%	15%	20%	25%	30%	35%	40%
5	-2.4	-2.2	-2.0	-1.6	-1.4	-1.1	-0.97	-0.83
10	-4.3	-3.8	-3.2	-2.5	-1.9	-1.5	-1.2	-1.00
15	-5.7	-4.9	-3.9	-2.9	-2.1	-1.7	-1.3	-1.10
20	-6.8	-5.6	-4.4	-3.1	-2.2	-1.7	-1.3	-1.10
25	-7.6	-6.2	-4.7	-3.1	-2.3	-1.7	-1.3	-1.10
30	-8.2	-6.5	-4.8	-3.2	-2.3	-1.7	-1.3	-1.10

Coefficient "c" (NPV of $1 of Working Capital Change)

	Discount Rates							
Lives	10%	12%	15%	20%	25%	30%	35%	40%
5	-0.28	-0.31	-0.33	-0.35	-0.36	-0.35	-0.33	-0.32
10	-0.56	-0.58	-0.60	-0.59	-0.55	-0.51	-0.47	-0.42
15	-0.77	-0.78	-0.76	-0.70	-0.63	-0.56	-0.50	-0.45
20	-0.92	-0.91	-0.86	-0.76	-0.66	-0.58	-0.51	-0.45
25	-1.0	-1.0	-0.92	-0.79	-0.68	-0.59	-0.52	-0.46
30	-1.1	-1.1	-0.95	-0.80	-0.68	-0.59	-0.52	-0.46

Coefficient "d" (NPV of $1 of One-time Cost)

	Discount Rates							
Lives	10%	12%	15%	20%	25%	30%	35%	40%
For Any Life	-0.60	-0.59	-0.57	-0.55	-0.52	-0.50	-0.49	-0.47

Table V-3 Shortcut NPV Equation Coefficients
Small Improvement Projects (<± $5 million)
Coefficient "a" (NPV of $1 of Investment)

Discount Rates

Lives	10%	12%	15%	20%	25%	30%	35%	40%
5	-0.83	-0.82	-0.81	-0.78	-0.76	-0.74	-0.72	-0.70
10	-0.93	-0.91	-0.88	-0.83	-0.80	-0.77	-0.74	-0.71
15	-1.0	-0.97	-0.92	-0.86	-0.81	-0.78	-0.75	-0.72
20	-1.1	-1.0	-0.95	-0.87	-0.82	-0.78	-0.75	-0.72
25	-1.1	-1.0	-0.96	-0.88	-0.82	-0.78	-0.75	-0.72
30	-1.1	-1.1	-0.97	-0.88	-0.82	-0.78	-0.75	-0.72

Coefficient "b" (NPV of $1 of Cost Minus Revenue)

Discount Rates

Lives	10%	12%	15%	20%	25%	30%	35%	40%
5	-2.6	-2.4	-2.2	-1.9	-1.6	-1.4	-1.3	-1.1
10	-4.5	-4.1	-3.5	-2.8	-2.3	-1.9	-1.6	-1.4
15	-6.0	-5.3	-4.4	-3.3	-2.6	-2.1	-1.7	-1.4
20	-7.2	-6.1	-4.9	-3.5	-2.7	-2.1	-1.7	-1.5
25	-8.0	-6.6	-5.2	-3.6	-2.7	-2.1	-1.8	-1.5
30	-8.7	-7.0	-5.4	-3.7	-2.7	-2.2	-1.8	-1.5

Coefficient "c" (NPV of $1 of Working Capital Change)

Discount Rates

Lives	10%	12%	15%	20%	25%	30%	35%	40%
5	-0.30	-0.33	-0.37	-0.41	-0.43	-0.44	-0.43	-0.43
10	-0.59	-0.63	-0.66	-0.68	-0.67	-0.64	-0.61	-0.57
15	-0.81	-0.84	-0.84	-0.81	-0.76	-0.71	-0.65	-0.60
20	-0.98	-0.98	-0.95	-0.88	-0.80	-0.73	-0.67	-0.61
25	-1.1	-1.1	-1.0	-0.91	-0.81	-0.73	-0.67	-0.61
30	-1.2	-1.1	-1.1	-0.92	-0.82	-0.74	-0.67	-0.61

Coefficient "d" (NPV of $1 of One-time Cost)

Discount Rates

Lives	10%	12%	15%	20%	25%	30%	35%	40%
For Any Life	-0.60	-0.59	-0.57	-0.55	-0.52	-0.50	-0.49	-0.47

Table V-4 Economic Assumptions by Project Type

→

Project Type	Large Project	Midsize Project	Small Improvement Project
Project Characteristics	New Process Area Entire System > ± $20 Million		Existing Process Area Improvements Pieces of Equipment ≤ ± $5 Million
Key Assumptions			
1. Years until startup	3	2	1
2. Project spendout (% years 1, 2, &3)	0, 40, 60	0, 100, 0	100, 0, 0
3. Startup costs (as % of proj. investment)	10	6	4
4. Annual creep investment (as % of proj. investment)	2	2	1
5. Annual maintenance (as % of proj. investment)	4	3	2
6. Yr. 1 operating cost/savings (as % of ongoing annual)	50	100	100
7. One-time cash cost (% years 1 & 2)	0, 100	100, 0	0,100
8. One-time working capital adjustment (% years 2, 3, 4, & 5)	0,0,50,50	0,100,0,0	100,0,0,0

Other assumptions common to all project types:
1. Annual costs and revenues are for a typical year.
2. All dollars are expressed in today's dollars.
3. U.S. tax and depreciation rates.
4. 4% escalation rate for all cash flows.
5. Automatic working capital change of 2 months of cash costs.
6. Only terminal value is working capital liquidation.
7. End of year cash flows.
8. 2% of investment for annual property taxes and overheads.

Adjustments to the NPV Estimations Due to Project Timing

Adjustments to the NPV calculated by the shortcut method are required if the actual project timing is different than that assumed for that project type. By way of example, assume that project startup occurs one year later than the assumptions shown in Table V-4 for that project type. In this case, the calculated NPV must be discounted by multiplying by 1/(1 + discount rate). If the discount rate is 25%, then the calculated NPV must be multiplied by 1/(1 + 0.25) or 0.80. If project startup occurs two years later, than the calculated NPV must be multiplied by $(1/(1 + 0.25))^2$ or 0.64. In this way, differences in project timing are made by either discounting or compounding the calculated NPV.

Comments on NPV Method

The sensitivity of the NPV to each of the assumptions listed in Table V-4 has been tested. With only a few exceptions, changing the assumption within reasonable limits does not have a significant impact on the calculations. One exception is the total of 4% to 6% of investment for annual maintenance, property taxes, and overheads. If this differs by more than two percentage points (for example, due to unusually high or low maintenance), there will be about a 10% error in the investment term. One way around this is to include an additional operating cost to account for maintenance above the default 2% to 4%.

Another important assumption is the 4% inflation rate. If inflation differs by one percentage point from this assumption, the cost and working capital change terms will have errors of around 8%.

The other assumption that is critical is in the working capital change term. The working capital is assumed to be liquidated at the end of the project, which has a large impact on the NPV of the working capital change. Note that this is not an important assumption for the cost term, which automatically includes a small amount of working capital.

It is important to note that the term for investment includes everything that is factored off of investment, including maintenance and other operating costs, as well as working capital change associated with the change in cost. The contribution of the pure investment to the NPV is only around 67% of the investment term, with around 6% attributable to the startup and project liaison costs, 9% due to creep investment, and 18% due to maintenance, property taxes and other costs factored from investment. On the other hand, although an automatic working capital change of two months of cash costs is included in the cost term, it only accounts for about 3% of the NPV of the cost term.

The accuracy of this shortcut method must always be considered. The calculations were done very accurately, but there is always uncertainty in the input data and the validity of the assumptions. Therefore, no more than two significant digits were used in the assumptions and the tabular output so as not to overstate the accuracy. Under the set of assumptions shown in Table V-4, the equation is exact. Under reasonably close assumptions, the error is likely to be less than 5-10%.

An Example

A pollution prevention alternative requires $5 million in new investment (small improvement project), expressed in today's dollars. Compared to the base case, it results in annual pretax savings of $2 million in disposal fees, expressed in today's dollars. What is the NPV for this alternative, compared to the base case?

First, the two primary cash flows are calculated. Investment (I) is $5 million. Savings (CS) are $2 million.

Assume for the moment that the appropriate life is 10 years and that the appropriate discount rate is 25%. This might be the case for a product which is nearing the end of its life cycle and has capital constraints. The NPV is then

$$NPV_{25} = -0.80 * I - (-2.3) * (CS + R)$$

$$= -0.80 * 5M - (-2.3) * (2M)$$

$$= + \$0.6M \text{ or } \$600,000$$

Opportunity Assessment

For each of the highest ranked ideas that were presented during the brainstorming session a champion will fill out an opportunity assessment form. This form contains the information needed to do a final ranking and to decide which ideas to pursue further. The form contains the following sections.

Current Operation

Describe the process with sufficient detail to identify the problem being addressed. The description should include the reason or driving force requiring improvements. If the process is not improved, describe what are the cost penalties.

Proposed Operation

The improvement being proposed has to be described in sufficient detail for another person to read and understand how to implement the improvement. Give some estimate on the cost to implement and any cost savings or additional revenues.

Assumptions

What are the assumptions that were used to believe that the new improvement would work. Describe any tests or studies that would be required to test the assumptions. Give an estimate of any unusual out-of-pocket costs required to develop and implement any tests or studies.

Advantages

Describe the advantages of the new operation over the current operation.

Concerns

Describe any areas of concern for the new operation. Include any guidance on how to overcome any areas of concern.

Technical Viability

Give your assessment of the technical viability. If the technical viability is medium or lower, the concerns should describe reasons why such a low rating.

Implementation

The three categories reflect the time required to implement.

- Procedural—The changes to be made to the process require minimal resources to implement and will not take a long period of time normally less than 6 months.

- Engineering Study—The process change requires engineering study, design and/or procurement of equipment and might take 6 to 12 months to implement.

- Research and Development—The process changes require a larger engineering or research study which could include lab or pilot plant testing and might take up to 2 to 5 years to complete.

Economic Benefits of Implementation

The economic benefits of the process change are summarized and consist of

- Investment

- Operating Costs Change

- Revenue Change

- Working Capital

- One Time Cost Changes

- Net Present Values at 10%, 25% and 40 % discount rates for a 10 year time period.

- On this summary page detail the changes to **inputs** to the process, that is, process feeds, solvents and others, and to **outputs** from the process, that is, products and wastes.

Calculation Worksheet

The calculation work sheet develops the numbers for the economic summary under the Economic Benefits of Implementation. For every number in the calculation worksheet there is a corresponding description in the various sections in the first part of the form. For example, if a solvent is eliminated from the process, then the Advantages section would have noted this fact, and the Materials and Working Capital sections of the worksheet will contain values.

Investment

Investment is the capital necessary to provide the installed equipment and all the auxiliaries that are needed for complete process operation. The factors that are used to calculate the investment reflect the state of the purchased equipment.

Bare Equipment

The purchased cost is for equipment that arrives on site in pieces and has to be assembled.

Skid Mounted

The purchased cost is for equipment that is modular and arrives on skids that are then attached together.

Vendor Installed

The purchased cost is for equipment installed by the vendor.

Project Level

The full project level investment has been estimated.

Operating costs

The operating costs are the full year costs or savings (a positive cost) for which the company pays with cash.

Materials

The materials costs are for any materials changes such as solvent use, feed materials, additives, catalysts, acids, bases, etc. The total material costs are in $1,000 per year.

Manpower

The manpower costs are for any additional manpower that has to be hired and overtime required to implement the improvement opportunity. Conversely any manpower savings (a positive cost) are for any person that can be removed from the company rolls or reduced overtime labor. The total values are in $1,000 per year.

Utilities

The utility cost changes are for changes in steam, electricity, brine, compressed air, nitrogen, process water, cooling tower water, chilled water, etc. The total values are in $1,000 per year.

Revenues

The change in revenues reflects the extra product produced or the changing of a waste to a valuable co-product. An increase in revenues is positive, and the total is in $1,000 per year.

Working capital change

Working capital is defined as the inventory change of material in the process. For example if a solvent is eliminated from the process, the working capital change (a positive value) would be the value of the average inventory, that is a half of a tank or some other amount. If a new catalyst is required, then the amount kept in inventory would be a working capital increase (a negative value). For the majority of the improvement ideas there is no working capital change. The value is in $1,000 per year.

One-time cost

Examples of one time costs are feasibility studies or tests done by vendors to validate the idea. Another one time cost would be where overtime is required to implement an idea. The value is in $1,000 per year.

Net present value (NPV)

NPV is the after-tax worth in today's dollars of all the future cash that an alternative will either consume or generate. NPV includes the effects of the four primary economic elements: investment, cash costs, revenues, and taxes. It is the most popular single measure of the economic merit of an alternative. The factors listed here are for 10 year period at 10%, 25% and 40% discount. Since most processes and products are changing rapidly, a period of 10 years was picked. Refer to Tables V-1 to V-3 for other time periods and discount rates.

Sketch of Current Approach

If there is a change in the physical arrangement of the equipment or if new equipment will be added, then a sketch of the present arrangement is required.

Sketch of Proposed Approach

Develop a sketch for the arrangement of any equipment needed to implement the improvement idea.

Revisit Opportunity Assessment

The original opportunity assessment was done on the third day of the brainstorming session. The first assessment is done to take advantage of

- The discussions that occurred the previous two days, and

- The diversity of the brainstorming team.

Before he/she leaves, each champion who fills out an opportunity assessment form needs to:

- Leave a copy with the core team leader,

- Take it back to the office and confirm the technical and economic aspects of the assessment, and

- Send an updated copy of the assessment form to the team leader.

At the discretion of the team leader, the brainstorming team could meet a second time one to two weeks later. At this meeting each champion would present a 5- to 15-minute summary of his or her assessment.

The team ranks the opportunities in order of implementation.

Final Report

The final report contains not only the results of the opportunity identification process, but also the information package sent to brainstorming participants. This report is the historical record of the program. With the rapid change in technology, feed materials, product requirements, environmental regulations and the normal drift of a manufacturing process from its optimum operating point, an opportunity identification program should be done every 3 to 5 years.

The parts of the report are:

- Cover Letter—The cover letter should identify the process, the overall waste reduction realized from the best ideas, any new investment required and the total net present value for a given discount rate and time period.

- Introduction—Identify the manufacturing process on which an opportunity identification program was done. Identify the main drivers and why this manufacturing process, contaminants and waste stream(s) were picked.

- Summary—Give a brief abstract of the highest ranked ideas that will be carried forward. Include a brief description of the idea and its value to the business along with any new investment.

- Top Ideas—List the top ideas by rank order. The list should contain the idea, waste reduction realized, any new investment, and NPV.

- Assessment Forms—Attach the assessment forms in order of value.

- List All Ideas—Develop a table of all ideas and include any descriptive material.

- Appendix—Attach a complete copy of the information package sent to the participants.

Appendix A

Forms and Handouts

Purpose

To discover the best ideas for process improvements requires:

- Sufficient information

- Diverse group of creative people

- Structured framework where the creative people can use the information to discover the best ideas.

This section contains a series of forms and handouts to:

- Identify waste stream(s) on which to focus creativity effort.

- Structure the process information for assimilation by a diverse group of creative people

- Pose a series of questions to start the creative thought process.

- Analyze and rank the best ideas.

Forms and Handouts In This Appendix

1) Waste Stream Description Form

2) Process Flow Chart for a Chemical Process

3) Function Description Form

4) Process Chemistry

5) Process Constituents and Sources

6) Component Property Form

7) Participants Responsibilities

8) Sample Participation Letter

9) Typical Questions for Each Participant to Consider

10) Typical Ground Rules for a Brainstorming Session

11) Brainstorming Facilitator, Scribe(s) and Process Expert(s) Duties

12) Opportunity Assessment Form

Waste Stream Description Form

Purpose

The purpose of this form is to gather sufficient information on the waste streams in the process to be able to

1) Rank the waste streams to identify the most important streams on which to focus.

2) Analyze the stream information to determine the source of the constituents in the waste stream. Once the sources are known then a series of creative ideas can be developed on how to eliminate or minimize the impact of the materials in the waste stream.

Information Required

The information to be entered on this form is:

- Date and identification information

- Component information ordered from the highest to the lowest rate.

- Each component is identified by name, per cent in the stream, whether or not it is on any regulatory list and the source of the material.

- Finally, the stream total flow and any other information about the stream such as pH, odor or any other characteristics that would make it a target stream.

The component amounts do not need to be precise, but an order of magnitude, for example, 0.1% versus 1% versus 10%. Most waste streams have a large flow constituent such as water and air and many minor constituents. The large flow is normally greater than 95% of the stream, and it controls the process and end-of-pipe treatment investment and operating costs. The minor constituents control the need to treat the stream.

Waste Stream Description

Date:
Process:
Waste Stream ID **Before** any Treatment Device:
Waste Description:
Waste Treatment (yes or no):
High Toxicity (yes or no):
Special Safety Hazard (yes or no):

Waste Compounds and Composition in order of Importance

Compound Name List in order from high to low flow rates	Wt% Vol% Mol%	Listed Yes or No**	Waste Origin that is where it is introduced to the process, or is introduced to the waste stream.

** Listed – on any regulatory lists

Stream Total Flow:
Units of Flow:
Other items of concern (odor, pH, etc.):

Process Flow Chart for a Chemical Process

Purpose

Experienced chemical engineers can read process flowsheets, follow the movement of products and other materials and understand the function of each major piece of equipment. However, some members of the opportunity identification team may not be trained to read and understand these flowsheets. To help those team members the process expert develops a process flow chart along with the function descriptions of the major process units depicted in the flowsheet.

Flow Chart Format

Chemical processes can be categorized into the following functions:

- Inputs and Outputs

- Chemical reactions.

- Heating and cooling systems.

- Separation systems.

- Flow movement devices such as pumps and compressors.

- Storage units.

Process lines connect these functions. For purposes of waste minimization the functions normally required to describe a process are—Inputs/Outputs, Heat/Cool, Reactions and Separations. Flow movement devices such as pumps are normally part of a separations or reactions system. The exception is a compressor because it is normally a large energy user. Storage units are in many cases part of the input/output, reactions and separation systems.

One format for a Process Flow Chart for a chemical process is to list the process flowsheet step in the left-hand column and have the functions listed along the top row. Examples are on the next page and in Appendix B.

Process Unit	Inputs/Outputs	Heat/Cool	Reactions	Separations
Catalytic Reactor	Feed		Reactor	
Distillation Column				Separate Solvents and Feeds
Distillation Column	Waste			Separate Product and Waste
	Product			

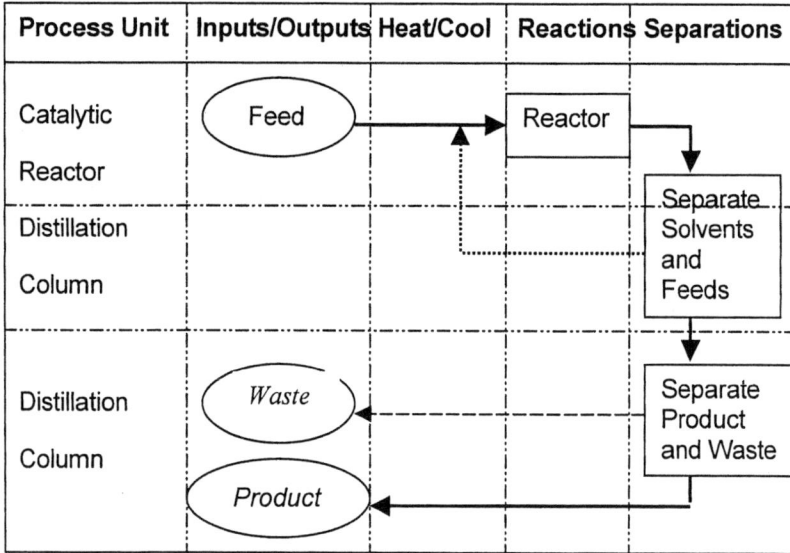

Process Flow Chart

Function Description Form

Purpose

The purpose of this form is to describe the function of each step in the flow chart for the team member who is not familiar with the process. The description contains information on <u>how</u> the particular process step operates and contains sufficient information to help identify changes to improve the present operation or modify the basic operation to realize a step change in the function's operation.

The function description form contains:

- Flowsheet Designation

- Type of Unit (reactor, distillation column, heater, tank, and so on)

- Function of Unit

- Principal Control Parameter(s)

- Principal Poor-operation Problems

- Uptime (Average time between outages)

- Wastes going to treatment or being emitted to the environment

Function Description Form

Flowsheet Designation:

Type of Unit (reactor, distillation column, heater, tank, and so on):

Function of Unit:

Principal Control Parameter(s):

Principal Poor-operation Problems:

Uptime (Average time between outages):

Wastes going to treatment or being emitted to the environment
       Process (total lbs./hr):
              Vapor or Gas:
              Liquid (non-aqueous):
              Water type:
              Solid:
              Hazardous:

Other
       Uptime losses (lbs. Generated at Startup and Shutdown):
       Maintenance losses:
       Storage losses:

Process Chemistry

The third necessary information to include in the data package is a description of the materials in the process, chemical reactions and reactor operating conditions. This includes

- Key physical and chemical properties for the feed materials, catalysts, byproducts, and products,

- Definition of the chemical reactions (sequential or parallel), particularly noting if the desired product is an intermediate in the reaction sequence,

- Acceptable reactor operating conditions,

- How reactants and products are added to and removed from the reactor,

- Trace impurities,

- Equilibrium relationships,

- etc.

Process Chemistry Form

Reactor Designation:

Chemical Reactions:

Key Reaction Parameters:

Order of Feed Material Addition:

Order of Product(s) Removal:

Solvents, Catalysts and Trace Materials:

Byproduct(s) or Waste(s):

Other Pertinent Information to Describe the Reactions and Reactor:

Process Constituents and Sources

Purpose

From a business perspective the only materials in the process that have business value are the low cost feed materials, intermediate(s) and salable product.. Other than the feed materials, the input of all other materials such as catalyst, solvents, water, air etc. are required because of the designers' limited knowledge of how to manufacture the product without them. The *function* of those material inputs to the process is to transform feed materials to products. To reduce the amount of non-value-adding materials being input to the process, either the feed materials, intermediates, or products must serve the same *function* as those other input materials, or the process needs to be modified to eliminate them. To help frame the problem for an existing manufacturing facility, then, a Process Analysis should be completed.

To do the Process Analysis described in Section III the constituents are listed in the following order in the Process Constituents and Sources form.

- Products

- Intermediates

- Feed Materials

- All others

The information in the Component Property Form plus the Process Analysis can help a creative person arrive at new basic process improvement ideas.

FORM: Process Constituents and Sources

Process: _____   Date: _____

List 1
_____

Constituent: Source (Feed, Reactor, Unit Operation):

Salable Products

_____   _____
_____   _____
_____   _____
                                   _____

Intermediates (result in salable products)

_____   _____
_____   _____
_____   _____
_____   _____
_____   _____
                                   _____

Essential Feed Materials(only those constituents used to produce the intermediates
and salable products)

_____   _____
_____   _____
_____   _____
_____   _____
_____   _____
                                   _____

List 2
_____

Constituent: Source (Feed, Reactor, Unit Operation):

Other Materials (Nonsalable byproducts, solvents, water, air, nitrogen, acids, etc.)

_____   _____
_____   _____
_____   _____
_____   _____
_____   _____
_____   _____
_____   _____
_____   _____

Component Information

Purpose

Using component information combined with the Waste Stream and Process Analyses a creative person is able to identify process improvements to reduce waste generation.

The minimum information for each component is:

- Name

- Formula

- Molecular weight

- Density

- Normal boiling point

- Normal freezing point

Special information such as solubility factors, activity coefficients, vapor pressures, etc. will be needed to a definitive analysis of the best ideas and is normally done off-line after the initial ranking.

Component Property Form

Name	Formula	Mol. Wt.	Density	T_b	T_s
Products					
Intermediates					
Feed Materials					
Others					

Density in Lb./Cu. Ft.
T_b Normal boiling point °C
T_s Normal freezing point °C

Participant Responsibilities

To maximize the contribution of the group, each individual group member needs to follow three ground rules—participate, be concise, and be additive.

First, each individual invited to the session is expected to participate—silence is unacceptable. All participants need to understand that their ideas, no matter how off-the-wall they may sound at first, will not be judged during the brainstorming portion of the meeting. The judgment and critique will be reserved for the screening portion of the session. The concept behind brainstorming is that one idea should lead to a new idea or build on the previous one. If a person does not speak up, then that individual's ingenuity is not being fully exercised.

Second, participants need to be concise. Ideas must be conveyed clearly and completely. Answer any questions on the meaning of your idea, but do not engineer the idea. There just is not enough time during the brainstorming portion of the meeting to engineer every idea. In addition, it will restrict the flow of new ideas.

Third, participants should be additive and avoid critiquing other people's ideas. Sometimes, an idea that was tried in the past and failed for either technological or political reasons will work in the current climate. Also keep in mind that all ideas will be reexamined at a later date. The goal of the brainstorming session is to get all possible ideas on the table, so that the best idea can be evaluated and chosen during the screening and evaluation stages of the assessment phase.

Outside Experts

Outside experts are invited to participate in the program to provide other viewpoints on how to improve the process. Thus, the experts have a special obligation to be prepared and additive. To be effective the outside expert should have:

- Reviewed the information in sufficient detail to understand the process,

- Developed a minimum of two improvement opportunities that address the basic root causes involving the chemistry, thermodynamics and engineering of the process,

- Communicated with the core team for any questions about the process.

An unprepared participant retards the brainstorming process by requiring the process expert to educate that person during the brainstorming session.

Process Engineers and Chemists

Process engineers and chemists who are participating in the brainstorming session have a special duty to be open-minded. If an idea is suggested that was tried in the past, the engineers or chemists cannot make any comment on the value of the idea. The fact that the idea was tried in the past and did not work is not relevant to the brainstorming. The idea could be the start for another participant to develop an even stronger idea. Also, the fact that the idea did not work in the past does not mean that the idea would not work now. **Thus, the process engineer and chemist must not say "We tried that idea in the past, and it did not work."**

Sample Invitation Letter

To ensure that the participants properly prepare for the brainstorming session, the invitation letter that accompanies the data package has to emphasize business leadership's expectations. Improper preparation by the participants could reduce the number of "good" ideas by 50%. Any idea not presented is a lost opportunity.

Dear Sir or Madam:

You are invited to participate in a brainstorming session. The purpose of the session is to reduce waste generation for the _____ process. The waste generation reduction will primarily be accomplished though identification of process improvements. These process improvements will also increase the revenues to the business.

Due to your area of expertise and knowledge you have been invited. However, to meet business's expectations you need to prepare for the brainstorming session. As part of that preparation the core team recommends that you:

- Review all of the material carefully.

- Develop two (2) ideas by using the Opportunity Assessment form included in the data package.

Those people who do not adequately prepare will impede the discovery of good process improvement opportunities and reduce the number of "good" ideas by as much as 50%.

We are looking forward to your participation.

Sincerely yours,

Typical Questions For Each Participant to Consider

Each participant is expected to bring a different perspective to all ideas that are generated. This is why it is so important to pick the right mix of people for the session. The concept is to have everyone be aware of the interdependency of any one idea on the whole, and how that idea can impact other ideas.

For the process engineer: What is the life of the present process? The present product? What is the competition doing that the group should know about?

For the chemist: What are the principal factors affecting yield, conversion, and selectivity? If the reaction is reversible, can byproducts be back-reacted to the incoming feed materials or converted to other useable products? If non-salable products are homologues of the reactants or intermediates, how can they be converted and recycled? What other catalysts are possible? If excess reactants or inerts are being used, ask why? If air, water, or a solvent are being used, ask why?

For the separations specialist: If an exit gas or water stream is being generated, what other separation techniques could be used to eliminate the stream? For trace levels of contaminants, how can the separation unit operations be improved? If large amounts of energy are required, what other separation technologies are applicable? If significant heating, followed by cooling, and then reheating takes place, what other combinations of unit operations can be used to minimize energy usage?

For the environmental specialist: What are the hazardous, carcinogenic, or toxic materials in the waste and product streams that require or could require further treatment? What are the present and future (5-10 years out) environmental laws that impact the waste from this process? What end-of-pipe technologies are appropriate?

For the engineering evaluator: For the current waste streams, what are the end-of-pipe treatment costs? What is the cost of waste generation for the current process?

For the energy specialist:: What are the opportunities to save energy in the process? What are the process-to-process energy exchange opportunities? What are the corporate energy goals?

For the lead operator and maintenance representative: What operating procedures are outdated or not followed? How does poor operation affect waste generation? How can startup, shutdown, and maintenance wastes be reduced?

Typical Ground Rules for a Brainstorming Session

The ground rules are designed to help control the atmosphere in the room during the session. The ground rules are an extension of the participant responsibilities and will be agreed upon by the participants at the beginning of the meeting. They should be posted in the meeting room and used to facilitate an effective and productive session.

Ground Rules

Participate

All ideas are good ideas.

Stay Focused

Keep the business needs and purpose of the brainstorming session in mind.

Build on Ideas

Use other people's ideas to create synergy.

Be Polite

Listen to understand; the person talking has the floor.

Be Positive

Work to sharpen the ideas being generated.

Opportunity Assessment

For each of the highest ranked ideas that were presented during the brainstorming session a champion will fill out an opportunity assessment form. This form contains the information needed to do a final ranking and to decide which ideas to pursue further. The form contains the following sections.

Current Operation

Describe the process with sufficient detail to identify the problem being addressed. The description should include the reason or driving force requiring improvements. If the process is not improved, describe what are the cost penalties.

Proposed Operation

The improvement being proposed has to be described in sufficient detail for another person to read and understand how to implement the improvement. Give some estimate on the cost to implement and any cost savings or additional revenues.

Assumptions

What are the assumptions that were used to believe that the new improvement would work. Describe any tests or studies that would be required to test the assumptions. Give an estimate of any unusual out-of-pocket costs required to develop and implement any tests or studies.

Advantages

Describe the advantages of the new operation over the current operation.

Concerns

Describe any areas of concern for the new operation. Include any guidance on how to overcome any areas of concern.

Technical Viability

Give your assessment of the technical viability. If the technical viability is medium or lower, the concerns should describe reasons why such a low rating.

Implementation

The three categories reflect the time required to implement.

- Procedural—The changes to be made to the process require minimal resources to implement and will not take a long period of time, up to 6 months.

- Engineering Study—The process change requires engineering study, design and/or procurement of equipment and might take 6 to 12 months to implement.

- Research and Development—The process changes require a larger engineering or research study which could include lab or pilot plant testing and might take up to 2 to 5 years to complete.

Economic Benefits of Implementation

The economic benefits of the process change are summarized and consist of

- Investment

- Operating Costs Change

- Revenue Change

- Working Capital

- One Time Cost Changes

- Net Present Values at 10%, 25% and 40 % discount rates for a 10 year time period.

- On this summary page detail the changes to **inputs** to the process, that is, process feeds, solvents and others, and to **outputs** from the process, that is, products and wastes.

Calculation Worksheet

The calculation work sheet develops the numbers for the economic summary under the Economic Benefits of Implementation. For every number in the calculation worksheet there is a corresponding description in the various sections in the first part of the form. For example, if a solvent is eliminated from the process, then the Advantages section would have noted this fact, and the Materials and Working Capital sections of the worksheet will contain values.

Investment

Investment is the capital necessary to provide the installed equipment and all the auxiliaries that are needed for complete process operation. The factors that are used to calculate the investment reflect the state of the purchased equipment.

Bare Equipment

The purchased cost is for equipment that arrives on site in pieces and has to be assembled.

Skid Mounted

The purchased cost is for equipment that is modular and arrives on skids that are then attached together.

Vendor Installed

The purchased cost is for equipment installed by the vendor.

Project Level

The full project level investment has been estimated.

Operating Costs

The operating costs are the full year costs or savings (a _positive_ cost) for which the company pays with cash.

Materials

The materials costs are for any materials changes such as solvent use, feed materials, additives, catalysts, acids, bases, etc. The total material costs are in $1,000 per year.

Manpower

The manpower costs are for any additional manpower that has to be hired and overtime required to implement the improvement opportunity. Conversely any manpower savings (a _positive_ cost) are for any person that can be removed from the company rolls or reduced overtime labor. The total values are in $1,000 per year.

Utilities

The utility cost changes are for changes in steam, electricity, brine, compressed air, nitrogen, process water, cooling tower water, chilled water, etc. The total values are in $1,000 per year.

Revenues

The change in revenues reflects the extra product produced or the changing of a waste to a valuable co-product. An increase in revenues is positive, and the total is in $1,000 per year.

Working Capital Change

Working capital is defined as the inventory change of material in the process. For example if a solvent is eliminated from the process, the working capital change (a positive value) would be the value of the average inventory, that is a half of a tank or some other amount. If a new catalyst is required, then the amount kept in inventory would be a working capital increase (a negative value). For the majority of the improvement ideas there is no working capital change. The value is in $1,000 per year.

One-time Cost

Examples of one-time costs are feasibility studies or tests done by vendors to validate the idea. Another one-time cost would be where overtime is required to implement an idea. The value is in $1,000 per year.

Net Present Value (NPV)

NPV is the after-tax worth in today's dollars of all the future cash that an alternative will either consume or generate. NPV includes the effects of the four primary economic elements: investment, cash costs, revenues, and taxes. It is the most popular single measure of the economic merit of an alternative. The factors listed here are for 10 year period at 10%, 25% and 40% discount. Since most processes and products are changing rapidly, a period of 10 years was picked. Refer to Tables V-1 to V-3 for other time periods and discount rates.

Sketch of Current Approach

If there is a change in the physical arrangement of the equipment or if new equipment will be added, then a sketch of the present arrangement is required.

Sketch of Proposed Approach

Develop a sketch for the arrangement of any equipment needed to implement the improvement idea.

OPPORTUNITY ASSESSMENT

Business

Opportunity (Idea) #   *Date:*   *Your Name:*

Opportunity (Idea) Title:

Opportunity Assessment Form—

Current Operation:

Proposed Operation:

Assumptions:

Advantages:

Concerns:

OPPORTUNITY ASSESSMENT

Business

*Opportunity (Idea) #*_____ *Date:*_____ *Your Name:*_____

*Opportunity (Idea) Title:*_____

Technical Viability: None__Very Low__Low__Medium___High___Very High___

Implementation: Procedural___Engineering Study___Research&Development___
Procedural—do in 6 months; Engineering Study—do in 12 months;
Research&Development—do in 2 to 5 years

Benefits of Implementation:

Change in Investment, Operating Costs, Revenues, and Net Present Value with Implementation of the Proposed Operation ($ 1,000)

Net New Installed Investment	$	_____
Net Reduction in Operating Costs	$	_____ /yr.
Net New Revenues	$	_____ /yr.
Reduction in Working Capital	$	_____
One Time Costs	$	_____

Net Present Value (10 years, 10% Discount)	$	_____
Net Present Value (10 years, 25% Discount)	$	_____
Net Present Value (10 years, 40% Discount)	$	_____

Change to Inputs (Process Feeds, Solvents, Catalysts, etc.) and Outputs (Water wastes, vapor or gas wastes, solid wastes, product(s))

Inputs:		Outputs:	
Process Feeds	_____Ton/yr.	Product(s)	_____Ton/yr.
Solvents	_____Ton/yr.	Wastes	_____Ton/yr.
Other	_____Ton/yr.	(Water Gas Solid)____ [Check appropriate media]	

OPPORTUNITY ASSESSMENT		
Business		Page 3 of 8

Opportunity (Idea) # _____ *Date:* _____ *Your Name:* _____

Opportunity (Idea) Title: _____

CALCULATION WORKSHEET

Investment In $1,000   = Estimate * Factor

Use **negative investment for return of an external sale**.

Factors to determine project level Investment		Type (Check type of
	Factor	estimate input below)
Bare equipment factor	8.0	_____
Skid mounted equipment factor	4.0	_____
Vendor installed·factor	2.0	_____
Project level factor	1.0	_____

Investment Estimate in $1,000   =   _____   Project Level   _____

(Estimate * Factor)

Operating Costs   Use dollars per year for the following

Positive numbers are decreased costs, negative numbers are increased costs.

Materials	Quantity Saved/yr.	Units	$/Unit	$ Saved/yr. ($1,000)
_____	_____	____	____	_____
_____	_____	____	____	_____
_____	_____	____	____	_____
_____	_____	____	____	_____
_____	_____	____	____	_____
_____	_____	____	____	_____

Total material cost Saved =   _____   ($1,000/yr.)

OPPORTUNITY ASSESSMENT

Business

Opportunity (Idea) #_____ Date:_____ Your Name:_____

Opportunity (Idea) Title: _____

CALCULATION WORKSHEET (cont.)

Manpower Change—Only include impact if headcount is added or reduced, or if overtime costs are increased decreased **(negative is an increase)**

	Amount	Factors	$ Saved/yr. ($1,000)
Contractor headcount reduction (Number)		$	
Employee headcount reduction (Number)		$	
Contractor overtime reduction (hours/year)		$	
Employee overtime reduction (hours/year)		$	
Total labor cost Saved	= _____ ($1,000/yr)		

Positive is a reduction, negative is an increase

Utilities	Amount	Units	$/Unit	$ Saved/yr. ($1,000)
_____	_____	____	____	_____
_____	_____	____	____	_____
_____	_____	____	____	_____
_____	_____	____	____	_____
_____	_____	____	____	_____
_____	_____	____	____	_____
Total Utility Cost Saved	= _____	($1,000/yr.)		

Total Operating Costs Saved	($1,000/yr)	
Materials	$ _____	
Manpower	$ _____	
Utilities	$ _____	
Total	$ _____	

OPPORTUNITY ASSESSMENT

Business

Opportunity (Idea) # _____ *Date:* _____ *Your Name:* _____

Opportunity (Idea) Title: _____

CALCULATION WORKSHEET (cont.)

Revenues

Positive is increased revenue, negative is decreased revenue.

Source of Revenue	Amount	Units	$/Unit	$ Earned/yr. ($1,000)
_____	_____	____	_____	_____
_____	_____	____	_____	_____
_____	_____	____	_____	_____
_____	_____	____	_____	_____
Total Earned =	_____		($1,000/yr.)	

Working Capital Change—Positive is a decreased cost, negative is an increased cost.

Material	Amount Change	Units	$/Unit	Saved / yr. ($1,000)
_____	_____	____	_____	_____
_____	_____	____	_____	_____
_____	_____	____	_____	_____
Total Working Capital Saved	= _____	($1,000)		

One Time Cost ($ 1,000) $ _____

Positive is increased cost, negative is decreased cost

Enter the one time costs needed for Research and Development, Pilot Plant, Engineering Study, etc. Do not include project team time costs unless these costs are recharged to the facility account.

OPPORTUNITY ASSESSMENT

Business

Opportunity (Idea) # _____ *Date:* _____ *Your Name:* _____

Opportunity (Idea) Title: _____

Net Present Value (NPV) for 10 Year Life

$$NPV = a * I - b * (CS + R) - c * WC + d * OTC$$

I = investment: **CS** = Cost Savings: **R** = Revenues: **WC** = Working Capital:
OTC = One Time Costs

Investment Amount		≤ $5 million	$5 to $20 million	≥ $20 million
Discount Rate 10%	*a*	-0.93	-0.99	-1.0
	b	-4.5	-4.3	-3.8
	c	-0.59	-0.56	-0.49
	d	-0.6	-0.6	-0.56
Discount Rate 25%	*a*	-0.80	-0.72	-0.67
	b	-2.3	-1.9	-1.4
	c	-0.67	-0.55	-0.41
	d	-0.52	-0.52	-0.44
Discount Rate 40%	*a*	-0.71	-0.56	-0.49
	b	-1.4	-1.0	-0.66
	c	-0.57	-0.42	-0.27
	d	-0.47	-0.47	-0.35

Discount Rate 10%

$$NPV_{10\%} = \underset{(a)}{\_\_\_} * \underset{(I)}{\_\_\_\_\_} - \underset{(b)}{\_\_\_} * \underset{(CS+R)}{\_\_\_\_\_} - \underset{(c)}{\_\_\_} * \underset{(WC)}{\_\_\_\_\_} + \underset{(d)}{\_\_\_} * \underset{(OTC)}{\_\_\_\_\_}$$

$NPV_{10\%} = _____$ **($ 1,000)**

Discount Rate 25%

$$NPV_{5\%} = \underset{(a)}{\_\_\_} * \underset{(I)}{\_\_\_\_\_} - \underset{(b)}{\_\_\_} * \underset{(CS+R)}{\_\_\_\_\_} - \underset{(c)}{\_\_\_} * \underset{(WC)}{\_\_\_\_\_} + \underset{(d)}{\_\_\_} * \underset{(OTC)}{\_\_\_\_\_}$$

$NPV_{25\%} = _____$ **($ 1,000)**

Discount Rate 40%

$$NPV_{40\%} = \underset{(a)}{\_\_\_} * \underset{(I)}{\_\_\_\_\_} - \underset{(b)}{\_\_\_} * \underset{(CS+R)}{\_\_\_\_\_} - \underset{(c)}{\_\_\_} * \underset{(WC)}{\_\_\_\_\_} + \underset{(d)}{\_\_\_} * \underset{(OTC)}{\_\_\_\_\_}$$

$NPV_{40\%} = _____$ **($ 1,000)**

OPPORTUNITY ASSESSMENT

Business

Opportunity (Idea) # _____ *Date:* _____ *Your Name:* _____

Opportunity (Idea) Title: _____

Sketch of Current Approach

National Center for Cleaner Production

OPPORTUNITY ASSESSMENT

Business

Page 8 of 8

Opportunity (Idea) #   *Date:*   *Your Name:*

Opportunity (Idea) Title:

Sketch of Proposed Approach

Appendix B

Chemical Plant Final Report

To: Businessperson

From: Project Champion

Date

Subject: Process Improvement Opportunity Identification for the Intermediates Manufacturing Facility.

The 40-year-old manufacturing process generates an unacceptable level of waste and is located within city limits. New local environmental regulations require that air and water emissions be reduced. Previous analysis determined that the regulations could be met by installing an expensive end-of-pipe gas emission treatment device and a new pre-treatment unit on the process wastewater (before being sent to the existing wastewater treatment system). The business wants to reduce the investment required to meet the new regulations. At the same time, the business is considering replacement of the existing plant with a new plant built at a different location, partly because of the increased public scrutiny due to the new regulations and the public's realization that the process contains and emits noxious compounds. Before reaching a decision, the business used a waste minimization methodology to identify process improvement opportunities. The purpose of the review was to identify opportunities that would:

- Reduce the end-of-pipe investment requirements to treat a toxic gas stream, and

- Reduce the nitrogen content in the outfall from the wastewater treatment facility.

Three opportunities were identified that would require less than $50,000 of investment for the existing plant. Implementation of all three would reduce the total investment needed to meet the new regulations by $1.4 million, save $500,000 per year in operating costs, and reduce waste water flow by 77%.

If the business decides to relocate, a combination of several new improvements results in a new design that would reduce

WASTE REDUCTION FOR CLEANER PRODUCTION
BUSINESS UNIT—INTERMEDIATES MANUFACTURE

- waste generation by 98%,

- benzene emissions by 100%,

- nitrogen in waste water by 98%,

- wastewater by 97%,

- energy usage by 85%,

- new required investment by $7.8 million, and

- $1.5 million / yr in operating costs.

Intermediate Manufacturing Cleaner Production Opportunity Identification

Introduction

The traditional approach to process design has been to first engineer the process and then engineer the treatment and disposal of waste streams. However, with increased regulatory and societal pressures to eliminate waste emissions to the environment, the total system must now be analyzed—process plus treatment—to find the most economic option. In general, processes that minimize the amount of waste generated at the source are the most economical. Even in existing facilities, waste generation can be reduced by more than 30% on average, while at the same time lowering operating costs and new capital investment.

This appendix presents a case study of a chemical manufacturing facility, which was faced with the task of reducing air and wastewater emissions. The case study illustrates both the "how to" of a waste minimization program as well as the tools and techniques that help engineers and scientists to identify process improvements that reduce waste generation inside the pipes and vessels. The case study is indicative of what can be done to reduce waste generation and emissions from existing manufacturing facilities. Given the right situation——significant strides can be made in protecting the environment, while making more money for the business.

Summary

Three of the top six ideas would require less than $50,000 of new capital investment and less than six months to complete in the existing plant. Two of the three short term projects would realize greater than $250,000 per year in savings and one would realizes greater than 70% reduction in waste generation.

These three opportunities all involve recycling of various outflows from the plant. The flowsheet for the existing plant is shown on page B-58.

The first opportunity is to recirculate 90% of the vent gas from the acid scrubber back to the compressor intake. This will require oxygen to be added to the compressor intake, and will not reduce waste contaminant flow to the new treatment device. However, reactant and product flow to the treatment device will be reduced. The major benefit is that by reducing the total flow to the treatment device, the size of the device is reduced significantly. This allows investment savings as well as savings in operating costs.

The second opportunity is very similar to the first. By recycling most of the discharge water from the steam stripper back to the water scrubber, the wastewater flow is significantly reduced. Even though the ammonia flow to pre-treatment will stay the same, the cost of installing and operating the pre-treatment system is reduced due to the lower total flow.

The third opportunity will install existing surplus equipment to condense the steam and benzene currently flowing to a thermal oxidizer. The benzene will be recycled back into the process, while the wastewater is sent to the existing treatment plant. Benzene purchases will drop nearly 300,000 pounds per year, but reduced operating costs at the thermal oxidizer provide most of the benefit.

If a new plant is built, the combining of several of the improvement ideas would reduce waste by more than 98%, benzene emissions by 100%, waste water by 97%, energy usage by 85%, new required investment by $7.8 million, and $1.5 million / yr. in operating costs. The combined process is shown in the following figure.

Pilot testing of a fluidized-bed reactor revealed that a 70% reduction in the total gas flow around the reactor was possible. The lower gas flow, coupled with the ability to maintain a high-pressure recirculating gas loop around the reactor, reduced compressor energy requirements by more than 85%. The fluidized bed and the use of make-up oxygen obviated the need for large quantities of air, resulting in significant savings in the gas compressor. Likewise the uniform temperature control of a fluidized bed and the exothermic heat of reaction eliminate the need for a process heat exchanger to preheat the feed. More importantly, reactant and product concentrations increased with the decreased airflow, allowing for direct condensation of the bulk of the product. Due to the boiling point differences between the reactant and product, a product-rich condensate is removed in this step (80%), bypassing the absorption step and part of the distillation train. The remainder of reactant and product are scrubbed using chilled reactant (i.e., reactant in place of water) in a considerably (60%) smaller-diameter unit.

The recirculating gas loop reduces by greater than 97% the nitrogen content from ammonia in water to wastewater treatment. The ammonia is recycled to the reactor and the slipstream to the absorber is less than 3% of the recirculating flow. The elimination of the water scrubber reduced dissolved nitrogen from NOx compounds in the gas stream.

The change in scrubbing medium has profound implications in further separations processing, in that the extraction step and raffinate stripping operations are completely eliminated. This substitution was arrived at in an effort to eliminate the large aqueous waste load imposed upon the process by aqueous scrubbing of moderately soluble organics. The elimination of the extraction step, benzene handling, and raffinate stripping operations provided significant project investment savings and lower operating costs. The equipment pieces and process/waste streams that were eliminated from and added to the redesigned process are

Equipment and Streams Eliminated (if a new plant is built)

- Reactor preheater

- Greatly reduced reactor feed gas compressor size

- Benzene extraction column

- Two absorption columns (scrubbers)

- Benzene handling facilities (tanks, pumps, etc.)

- Much smaller thermal oxidizer

- No water feed

- No dilute acid feed

Improved process figure

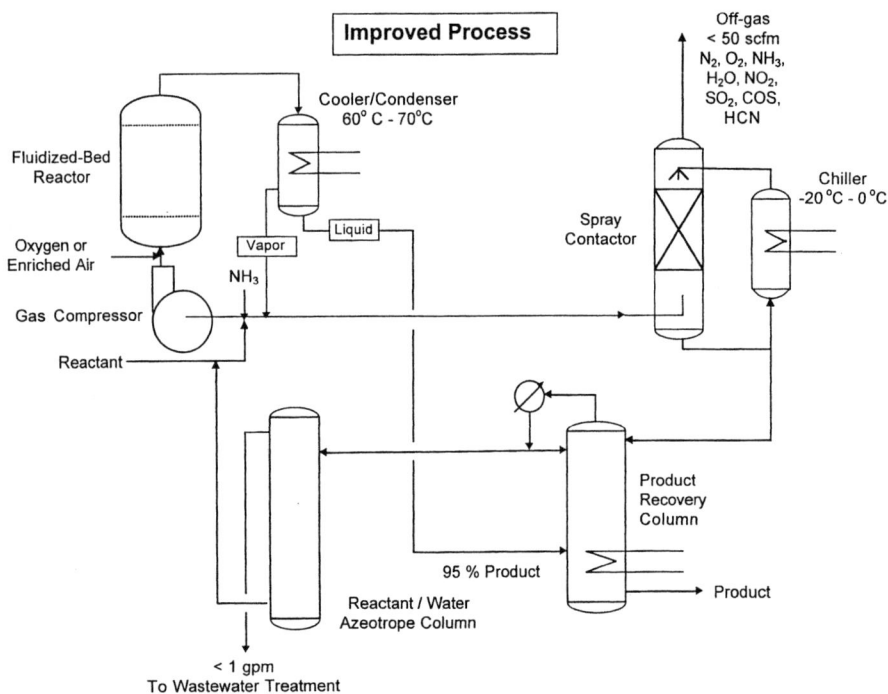

Improved Process

Off-gas
< 50 scfm
N_2, O_2, NH_3,
H_2O, NO_2,
SO_2, COS,
HCN

Cooler/Condenser
60° C - 70°C

Fluidized-Bed
Reactor

Chiller
$-20\,^\circ$C - $0\,^\circ$C

Spray
Contactor

Vapor   Liquid

Oxygen or
Enriched Air

NH_3

Gas Compressor

Reactant

Product
Recovery
Column

95 % Product

Product

Reactant / Water
Azeotrope Column

< 1 gpm
To Wastewater Treatment

Equipment Added (if a new plant is built)

- Product cooler/condenser downstream of reactor

- Spray contactor and chiller using cold REAC as absorbent

- REAC/water azeotropic recovery column to separate water of reaction

- Increased size of product recovery column

Emission reductions include:

- Vent gas flow was reduced from 10,000 scfm to less than 50 scfm;

- Wastewater flow was reduced from 35 gal/min to 1 gal/min; and

- All benzene fugitive emissions were eliminated.

In addition, thermal oxidizer emissions were lowered by more than 90% due to the smaller vent gas flow, and energy consumption was significantly reduced. Personnel exposure problems and administrative controls required with benzene are also no longer an issue with the revised process.

Changes in the distillation train are required; however, no measurable increase in capital investment results. In addition, the overall distillation train has a lower operating cost now that the benzene solvent recovery step is eliminated. The thermal oxidizer required to treat off-gases is also greatly reduced in size and cost. The following table details the financial benefits of the revised manufacturing process.

Financial Results
New Plant Design Vs Current Plant Design

	Cost or (Savings) $ Millions
Compressor rating reduced from 900 hp to 50 hp	(0.9)
Extraction equipment eliminated	(2.1)
Scrubber size reduced	(0.2)
Benzene handling facilities eliminated	(3.1)
Stream stripper eliminated	(0.4)
Thermal oxidizer size reduced	(1.6)
Product condenser required	0.5
Investment savings	7.8
NPV of savings (12%, 10 yr.)	12.9

Top Ideas

1) Recycle vent stream and add additional oxygen or enriched air as needed—New investment $50,000, cost savings $110,000 and 6 months to implement.

2) Recycle water exiting steam stripper to water scrubber—New investment $35,000, cost savings $300,000 and 6 months to implement.

3) Condense stripper steam and recover benzene—New investment $50,000, cost savings $300,000 and 6 months to implement.

4) Fluidized-bed reactor—If a new manufacturing facility is built, there is no increase in new investment. To replace the existing reactor at the existing plant would require an investment of $900,000 and would result in a $800,000 decrease in the waste gas treatment investment. The cost savings are $80,000/yr. and installation would require 2 to 3 years to implement.

5) Replace water scrubber with solvent scrubber—If a new plant is built, this reduces the net investment by $600,000, and increases operating costs by $220,000/yr. To install in the existing plant, the investment would be $2,100,000. It would take 1 to 2 years to implement.

6) Install condenser to remove product—If a new plant is built, there would be an improved product recovery of $48,000, that requires a new investment of $500,000, and an increase in operating costs of $30,000. To install in the existing plant, the investment would be $1,100,000 and would take 1 to 2 years to implement.

Impact of Ideas on Waste from Affected Unit Operations

Ideas	Benzene Emissions	Chemical Emissions	Waste	Waste Water	Energy
1. Recycle Vent Gas	None	20% reduction due to 90% reduction in lost reactant and product through vent	90% reduction in vent gas	None	2% compressor energy usage increase to move vent gas to compressor suction
2. Recycle Water	None	Insignificant	None	86% reduced	None – existing pump used
3. Condense Stripper Steam	10% reduction in benzene emissions	Insignificant	100% reduction in steam to TOX	9% increase due to condensed steam. Net 77% reduction with Idea 2.	Fuel gas usage is reduced significantly
4. Fluidized Bed	None	None if option 1 implemented	None if option 1 implemented	None	70% reduction in compressor energy use
5. Solvent Scrubber	100% reduction in benzene emissions	Insignificant	100% reduction in contaminated steam sent to TOX	97% reduced	100% reduction in steam usage. Increased electric for chiller
6. Non-contact Condenser	None	95% reduction in product loss through vent	None	Insignificant	None
Combined ideas 1, 4, 5, & 6 on total plant waste	100% reduction	90% reduction	98% reduction	97% reduction	85% reduction

OPPORTUNITY ASSESSMENT
Business

Opportunity (Idea) # 9   Date: Your Name:

Opportunity (Idea) Title: Recycle Vent Gas to Reactor

Opportunity Assessment Form— Recycle Vent Gas to Reactor

Current Operation:
The vent gas from the acid gas scrubber is emitted to the atmosphere. The 10,000 scfm vent gas contains noxious compounds such as COS and HCN.

Proposed Operation:
Recycle the vent back to the inlet to the compressor and add pure oxygen or enriched air as oxygen makeup. Investment required $50,000 and time to install is 6 months.

Assumptions:
The recycle of the by-products does not affect the operation of the catalyst. Also, the high water load from the scrubbers does not adversely affect the catalytic action.

Advantages:
The gas to the thermal oxidizer is reduced to 10% of the original flow. The thermal oxidizer investment is decreased by $1,150,000 and the operating cost by $300,000.

Concerns:
Possible adverse impact on the catalyst at the higher concentration of contaminants.

National Center for Cleaner Production

OPPORTUNITY ASSESSMENT
Business _____

Opportunity (Idea) # 9   Date: _____   Your Name: _____

Opportunity (Idea) Title: Recycle Vent Gas to Reactor

Technical Viability: None__ Very Low__ Low___ Medium___ High___ Very High__ **X**_

Implementation: Procedural___ Engineering Study_**X**__
          Research&Development___
Procedural—do in 6 months; Engineering Study—do in 12 months;
Research&Development—do in 2 to 3 years

Benefits of Implementation:

Change in Investment, Operating Costs, Revenues, and Net Present Value with Implementation of the Proposed Operation ($ 1,000)	
Net New Installed Investment	$ -1100
Net Reduction in Operating Costs	$ 110 /yr.
Net New Revenues	$ _____ /yr.
Reduction in Working	$ _____
One Time Costs	$ _____
Net Present Value (10 years, 10% Discount)	$ 530
Net Present Value (10 years, 25% Discount)	$ 630
Net Present Value (10 years, 40% Discount)	$ 630

Change to Inputs (Process Feeds, Solvents, Catalysts, etc.) and Outputs (Water wastes, vapor or gas wastes, solid wastes, product(s))			
Inputs:		**Outputs:**	
Process Feeds _____ Ton/yr.		Product(s) _____ Ton/yr.	
Solvents _____ Ton/yr.		Wastes 166610 Ton/yr.	
Other _____ Ton/yr.		(Water___ Gas X___ Solid___) [Check appropriate media]	

OPPORTUNITY ASSESSMENT
Business
_____

Opportunity (Idea) # 9   _Date:_ _____   _Your Name:_ _____

Opportunity (Idea) Title: Recycle Vent Gas to Reactor
_____

CALCULATION WORKSHEET

Investment In $1,000   = Estimate * Factor

Use **negative investment for return of an external sale**.

Factors to determine project level Investment

	Factor	Type (Check type of estimate input below)
Bare equipment factor	8.0	
Skid mounted equipment factor	4.0	_____
Vendor installed factor	2.0	_____
Project level factor	1.0	X

Investment Estimate in $1,000   =   -1150   Project Level   -1150

(Estimate * Factor)

Operating Costs   Use dollars per year for the following

Positive numbers are decreased costs, negative numbers are increased costs.

Materials	Quantity Saved/yr.	Units	$/Unit	$ Saved/yr. ($1,000)
Oxygen				-500
Thermal oxidizer operation				390
Total material cost Saved	=	-118		($1,000/yr.)

OPPORTUNITY ASSESSMENT
Business

Opportunity (Idea) # 9   *Date:*   *Your Name:*

Opportunity (Idea) Title: Recycle Vent Gas to Reactor

CALCULATION WORKSHEET (cont.)

Manpower Change—Only include impact if headcount is added or reduced, or if overtime costs are increased decreased   **(negative is an increase)**

	Amount	Factors	$ Saved/yr. ($1,000)
Contractor headcount reduction (Number)		$52,000/yr.	
Employee headcount reduction (Number)		$160,000/yr.	
Contractor overtime reduction (hours/year)		$35 /hr.	
Employee overtime reduction (hours/year)		$35 /hr.	

Total labor cost Saved   =_____ **($1,000/yr)**

• •

Positive is a reduction, negative is an increase

Utilities	Amount	Units	$/Unit	$ Saved/yr. ($1,000)
> 400 psig steam		1000# /yr.	3.75	
235 psig steam		1000# /yr.	2.93	
30 psig steam		1000# /yr.	1.84	
Electricity		KWH/yr.	0.03	
Air		MSCF/yr.	0.11	
Nitrogen		MSCF/yr.	0.39	

Total Utility Cost Saved   =_____ **($1,000/yr.)**

• ▪▪▪▪▪▪▪▪▪▪▪▪▪▪▪▪▪▪▪ • • • • • • • • •

Total Operating Costs Saved	**($1,000/yr)**
Materials	$ **-110**_____
Manpower	$ _____
Utilities	$ _____
Total	$ **-110**_____

OPPORTUNITY ASSESSMENT
Business

Opportunity (Idea) # 9   *Date:*   *Your Name:*

Opportunity (Idea) Title: Recycle Vent Gas to Reactor

CALCULATION WORKSHEET (cont.)

Revenues
Positive is increased revenue, negative is decreased revenue.

Source of Revenue	Amount	Units	$/Unit	$ Earned/yr. ($1,000)
_____	_____	_____	_____	_____
_____	_____	_____	_____	_____
_____	_____	_____	_____	_____
_____	_____	_____	_____	_____
Total Earned =	_____	($1,000/yr.)		

Working Capital Change—Positive is a decreased cost, negative is an increased cost.

Material	Amount Change	Units	$/Unit	Saved / yr. ($1,000)
_____	_____	_____	_____	_____
_____	_____	_____	_____	_____
_____	_____	_____	_____	_____
Total Working Capital Saved =	_____	($1,000)		

One Time Cost ($ 1,000) $ _____
Positive is increased cost, negative is decreased cost
Enter the one time costs needed for Research and Development, Pilot Plant, Engineering Study, etc. Do not include project team time costs unless these costs are recharged to the facility account.

OPPORTUNITY ASSESSMENT

Business _____

Opportunity (Idea) # __9__   *Date:* _____   *Your Name:* _____

Opportunity (Idea) Title: Recycle Vent Gas to Reactor _____

Net Present Value (NPV) for 10 Year Life

$$NPV = a * I - b * (CS + R) - c * WC + d * OTC$$

I = investment: **CS** = Cost Savings: **R** = Revenues: **WC** = Working Capital:
OTC = One Time Costs

Investment Amount		≤ $5 million	$5 to $20 million	≥ $20 million
Discount Rate 10%	a	-0.93	-0.99	-1.0
	b	-4.5	-4.3	-3.8
	c	-0.59	-0.56	-0.49
	d	-0.6	-0.6	-0.56
Discount Rate 25%	a	-0.80	-0.72	-0.67
	b	-2.3	-1.9	-1.4
	c	-0.67	-0.55	-0.41
	d	-0.52	-0.52	-0.44
Discount Rate 40%	a	-0.71	-0.56	-0.49
	b	-1.4	-1.0	-0.66
	c	-0.57	-0.42	-0.27
	d	-0.47	-0.47	-0.35

Discount Rate 10%

$$NPV_{10\%} = \underset{(a)\quad(I)}{\underline{0.93* -1100}} - \underset{(b)\quad(CS+R)}{\underline{-4.5 \ * \ -110}} - \underset{(c)\quad(WC)}{\underline{\quad * \quad}} + \underset{(d)\quad(OTC)}{\underline{\quad * \quad}}$$

$NPV_{10\%} = $ __530__ (\$ 1,000)

Discount Rate 25%

$$NPV_{5\%} = \underset{(a)\quad(I)}{\underline{-0.80 \ * -1100}} - \underset{(b)\quad(CS+R)}{\underline{-2.3 \ * \ -110}} - \underset{(c)\quad(WC)}{\underline{\quad * \quad}} + \underset{(d)\quad(OTC)}{\underline{\quad * \quad}}$$

$NPV_{25\%} = $ __630__ (\$ 1,000)

Discount Rate 40%

$$NPV_{40\%} = \underset{(a)\quad(I)}{\underline{-0.71 \ * -1100}} - \underset{(b)\quad(CS+R)}{\underline{-1.4 \ * \ -110}} - \underset{(c)\quad(WC)}{\underline{\quad * \quad}} + \underset{(d)\quad(OTC)}{\underline{\quad * \quad}}$$

$NPV_{40\%} = $ __630__ (\$ 1,000)

OPPORTUNITY ASSESSMENT
Business

Opportunity (Idea) # 36   *Date:*   *Your Name:*

Opportunity (Idea) Title: Recycle Water From Steam Stripper to Water Scrubber

Assessment Form— Recycle Water From Steam Stripper to Water Scrubber

Current Operation:
The water leaving the stream stripper is sent to wastewater treatment. The present Nitrogen content is too high, thus the stream will require pretreatment to remove the nitrogen containing compounds, investment $600,000.

Proposed Operation:
Recycle 30 gpm of the water to water scrubber to provide the water needed for that scrubber. Requires an investment of $35,000 to implement.

Assumptions:
Salts in the water do not impact scrubber operation.

Advantages:
Reduce effluent to pretreatment device from 35 to 5 gpm. Reduce investment of device to $200,000.

Concerns:
Level of nitrogen compounds to wastewater treatment does not change.

National Center for Cleaner Production

OPPORTUNITY ASSESSMENT

Business _____

Opportunity (Idea) # _36___   Date: _____   Your Name: _____

Opportunity (Idea) Title: Recycle Water From Steam Stripper to Water Scrubber

Technical Viability: None__ Very Low__ Low___ Medium___ High__ **X** _Very High___

Implementation: Procedural___ Engineering Study_**X**_ Research&Development___
Procedural—do in 6 months; Engineering Study—do in 12 months;
Research&Development—do in 2 to 3 years

Benefits of Implementation:

Change in Investment, Operating Costs, Revenues, and Net Present Value with Implementation of the Proposed Operation ($ 1,000)		
Net New Installed Investment	$	-365
Net Reduction in Operating Costs	$	300 /yr.
Net New Revenues	$	____ /yr.
Reduction in Working Capital	$	____
One Time Costs	$	____
Net Present Value (10 years, 10% Discount)	$	1690
Net Present Value (10 years, 25% Discount)	$	980
Net Present Value (10 years, 40% Discount)	$	680

Change to Inputs (Process Feeds, Solvents, Catalysts, etc.) and Outputs (Water wastes, vapor or gas wastes, solid wastes, product(s))			
Inputs:		**Outputs:**	
Process Feeds _____ Ton/yr.		Product(s) _____ Ton/yr.	
Solvents _____ Ton/yr.		Wastes 59130 Ton/yr.	
Other _____ Ton/yr.		(Water X Gas ___ Solid ___) [Check appropriate media]	

OPPORTUNITY ASSESSMENT
Business

Opportunity (Idea) # 36   Date: Your Name:

Opportunity (Idea) Title: Recycle Water From Steam Stripper to Water Scrubber

CALCULATION WORKSHEET

Investment In $1,000   = Estimate * Factor
Use **negative investment for return of an external sale**.

Factors to determine project level Investment

	Factor	Type (Check type of estimate input below)
Bare equipment factor	8.0	_____
Skid mounted equipment factor	4.0	_____
Vendor installed factor	2.0	
Project level factor	1.0	X

Investment Estimate in $1,000   =   -365   Project Level   -365

(Estimate * Factor)

Operating Costs   Use dollars per year for the following
Positive numbers are decreased costs, negative numbers are increased costs.

Materials	Quantity Saved/yr.	Units	$/Unit	$ Saved/yr. ($1,000)
Treatment costs				300
Total material cost Saved	=	300		**($1,000/yr.)**

OPPORTUNITY ASSESSMENT
Business _____

Opportunity (Idea) # 36 _____ _Date:_ _____ _Your Name:_ _____

Opportunity (Idea) Title: Recycle Water From Steam Stripper to Water Scrubber

CALCULATION WORKSHEET (cont.)

Manpower Change—Only include impact if headcount is added or reduced, or if overtime costs are increased decreased **(negative is an increase)**

	Amount	Factors	$ Saved/yr. ($1,000)
Contractor headcount reduction (Number)		$52,000/yr.	
Employee headcount reduction (Number)		$160,000/yr.	
Contractor overtime reduction (hours/year)		$35 /hr.	
Employee overtime reduction (hours/year)		$35 /hr.	

Total labor cost Saved   = _____ **($1,000/yr)**

• •

Positive is a reduction, negative is an increase

Utilities	Amount	Units	$/Unit	$ Saved/yr. ($1,000)
> 400 psig steam		1000# /yr.	3.75	
235 psig steam		1000# /yr.	2.93	
30 psig steam		1000# /yr.	1.84	
Electricity		KWH/yr.	0.03	
Air		MSCF/yr.	0.11	
Nitrogen		MSCF/yr.	0.39	

Total Utility Cost Saved   = _____ **($1,000/yr.)**

• ▪ ▪ ▪ ▪ ▪ ▪ ▪ • • • • • • • • • • • • • • • •

Total Operating Costs Saved	**($1,000/yr)**
Materials	$ 300 _____
Manpower	$ _____
Utilities	$ _____
Total	$ 300 _____

OPPORTUNITY ASSESSMENT
Business

Opportunity (Idea) # 36   *Date:*   *Your Name:*

Opportunity (Idea) Title: Recycle Water From Steam Stripper to Water Scrubber

CALCULATION WORKSHEET (cont.)

Revenues

Positive is increased revenue, negative is decreased revenue.

Source of Revenue	Amount	Units	$/Unit	$ Earned/yr. ($1,000)

Total Earned   =   ($1,000/yr.)

Working Capital Change—Positive is a decreased cost, negative is an increased cost.

Material	Amount Change	Units	$/Unit	Saved / yr. ($1,000)

Total Working Capital Saved   =   ($1,000)

One Time Cost ($ 1,000) $

Positive is increased cost, negative is decreased cost

Enter the one time costs needed for Research and Development, Pilot Plant, Engineering Study, etc. Do not include project team time costs unless these costs are recharged to the facility account.

National Center for Cleaner Production

OPPORTUNITY ASSESSMENT

Business

Opportunity (Idea) # **36**   *Date:*   *Your Name:*

Opportunity (Idea) Title: Recycle Water From Steam Stripper to Water Scrubber

Net Present Value (NPV) for 10 Year Life

$$NPV = a * I - b * (CS + R) - c * WC + d * OTC$$

I = investment: **CS** = Cost Savings: **R** = Revenues: **WC** = Working Capital:
OTC = One Time Costs

Investment Amount		≤ $5 million	$5 to $20 million	≥ $20 million
Discount Rate 10%	a	-0.93	-0.99	-1.0
	b	-4.5	-4.3	-3.8
	c	-0.59	-0.56	-0.49
	d	-0.6	-0.6	-0.56
Discount Rate 25%	a	-0.80	-0.72	-0.67
	b	-2.3	-1.9	-1.4
	c	-0.67	-0.55	-0.41
	d	-0.52	-0.52	-0.44
Discount Rate 40%	a	-0.71	-0.56	-0.49
	b	-1.4	-1.0	-0.66
	c	-0.57	-0.42	-0.27
	d	-0.47	-0.47	-0.35

Discount Rate 10%
$$NPV_{10\%} = \underline{\ 0.93^* \ } \underline{\ -365\ } - \underline{\ -4.5\ } * \underline{\ 300\ } - \underline{\quad} * \underline{\quad} + \underline{\quad} * \underline{\quad}$$
$$\qquad\qquad (a) \quad (I) \qquad (b) \ (CS+R) \quad (c) \ (WC) \ (d) \ (OTC)$$

$$NPV_{10\%} = \underline{\ 1690\ } \ (\$ \ 1{,}000)$$

Discount Rate 25%
$$NPV_{5\%} = \underline{\ -0.80\ } * \underline{\ -365\ } - \underline{\ -2.3\ } * \underline{\ 300\ } - \underline{\quad} * \underline{\quad} + \underline{\quad} * \underline{\quad}$$
$$\qquad\qquad (a) \quad (I) \qquad (b) \ (CS+R) \quad (c) \ (WC) \ (d) \ (OTC)$$

$$NPV_{25\%} = \underline{\ 980\ } \ (\$ \ 1{,}000)$$

Discount Rate 40%
$$NPV_{40\%} = \underline{\ -0.71\ } * \underline{\ -365\ } - \underline{\ -1.4\ } * \underline{\ 300\ } - \underline{\quad} * \underline{\quad} + \underline{\quad} * \underline{\quad}$$
$$\qquad\qquad (a) \quad (I) \qquad (b) \ (CS+R) \quad (c) \ (WC) \ (d) \ (OTC)$$

$$NPV_{40\%} = \underline{\ 680\ } \ (\$ \ 1{,}000)$$

OPPORTUNITY ASSESSMENT
Business

Opportunity (Idea) # **44**   *Date:*   *Your Name:*

Opportunity (Idea) Title: Condense stripper steam and recover benzene

Opportunity Assessment Form—Condense stripper steam and recover benzene

Current Operation:
The steam plus benzene being stripped from the wastewater is sent to a thermal oxidizer to be burned.

Proposed Operation:
Use an existing condenser plus decanter to condense the steam, separate the benzene and recycle the benzene back to the benzene extractor.

Assumptions:
The recycled benzene does not contain any contaminants to affect the extraction step.

Advantages:
Recover 33 lb./hr of benzene.

Concerns:
No particular disadvantages.

National Center for Cleaner Production

OPPORTUNITY ASSESSMENT

Business _____

Opportunity (Idea) # **44**   *Date:* _____   *Your Name:* _____

Opportunity (Idea) Title: Condense stripper steam and recover benzene _____

Technical Viability: None__Very Low__Low___Medium___High_**XX**Very High___

Implementation: Procedural___Engineering Study_**XX**Research&Development___
Procedural—do in 6 months; Engineering Study—do in 12 months;
Research&Development—do in 2 to 3 years

Benefits of Implementation:

Change in Investment, Operating Costs, Revenues, and Net Present Value with Implementation of the Proposed Operation ($ 1,000)	
Net New Installed Investment	$ 50
Net Reduction in Operating Costs	$ 108 /yr.
Net New Revenues	$ /yr.
Reduction in Working	$
One Time Costs	$
Net Present Value (10 years, 10% Discount)	$ 440
Net Present Value (10 years, 25% Discount)	$ 210
Net Present Value (10 years, 40% Discount)	$ 115

Change to Inputs (Process Feeds, Solvents, Catalysts, etc.) and Outputs (Water wastes, vapor or gas wastes, solid wastes, product(s))			
Inputs:		**Outputs:**	
Process Feeds	____ Ton/yr.	Product(s)	____ Ton/yr.
Solvents	109 Ton/yr.	Wastes	4941 Ton/yr.
Other	____ Ton/yr.	(Water___Gas X___ Solid___) [Check appropriate media]	

OPPORTUNITY ASSESSMENT
Business

Opportunity (Idea) # __44__ *Date:* _____ *Your Name:* _____

Opportunity (Idea) Title: Condense stripper steam and recover benzene

CALCULATION WORKSHEET

Investment In $1,000   = Estimate * Factor
Use **negative investment for return of an external sale**.
Factors to determine project level Investment

	Factor	Type (Check type of estimate input below)
Bare equipment factor	8.0	____
Skid mounted equipment factor	4.0	____
Vendor installed factor	2.0	
Project level factor	1.0	X

Investment Estimate in $1,000   =   50   Project Level   50

(Estimate * Factor)

Operating Costs   Use dollars per year for the following
Positive numbers are decreased costs, negative numbers are increased costs.

Materials	Quantity Saved/yr.	Units	$/Unit	$ Saved/yr. ($1,000)
Reduced thermal oxidizer costs				70
Recovered benzene	290,000	lbs./yr.	0.13/lbs.	38

Total material cost Saved =   **108**   ($1,000/yr.)

OPPORTUNITY ASSESSMENT

Business _____

Opportunity (Idea) # __44__   *Date:* _____   *Your Name:* _____

Opportunity (Idea) Title: Condense stripper steam and recover benzene _____

CALCULATION WORKSHEET (cont.)

Manpower Change—Only include impact if headcount is added or reduced, or if overtime costs are increased decreased   **(negative is an increase)**

	Amount	Factors	$ Saved/yr. ($1,000)
Contractor headcount reduction (Number)	$52,000/yr.		
Employee headcount reduction (Number)	$160,000/yr.		
Contractor overtime reduction (hours/year)	$35 /hr.		
Employee overtime reduction (hours/year)	$35 /hr.		

Total labor cost Saved   = _____ **($1,000/yr)**

■ ■

Positive is a reduction, negative is an increase

Utilities	Amount	Units	$/Unit	$ Saved/yr. ($1,000)
> 400 psig steam		1000# /yr.	3.75	
235 psig steam		1000# /yr.	2.93	
30 psig steam		1000# /yr.	1.84	
Electricity		KWH/yr.	0.03	
Air		MSCF/yr.	0.11	
Nitrogen		MSCF/yr.	0.39	

Total Utility Cost Saved   = _____ **($1,000/yr.)**

■ ■

Total Operating Costs Saved	**($1,000/yr)**
Materials	$ 108 _____
Manpower	$ _____
Utilities	$ _____
Total	$ 108 _____

OPPORTUNITY ASSESSMENT
Business

Opportunity (Idea) # **44**   Date: Your Name:

Opportunity (Idea) Title: *Condense stripper steam and recover benzene*

CALCULATION WORKSHEET (cont.)

Revenues

Positive is increased **revenue**, negative is decreased revenue.

Source of Revenue	Amount	Units	$/Unit	$ Earned/yr. ($1,000)
_____	_____	____	_____	_____
_____	_____	____	_____	_____
_____	_____	____	_____	_____
_____	_____	____	_____	_____
Total Earned	**=**		**($1,000/yr.)**	

Working Capital Change—Positive is a decreased cost, negative is an increased cost.

Material	Amount Change	Units	$/Unit	Saved / yr. ($1,000)
_____	_____	____	_____	_____
_____	_____	____	_____	_____
_____	_____	____	_____	_____
Total Working Capital Saved		**=** ____	**($1,000)**	

One Time Cost ($ 1,000) $ _____

Positive is increased **cost**, negative is decreased cost

Enter the one time costs needed for Research and Development, Pilot Plant, Engineering Study, etc. Do not include project team time costs unless these costs are recharged to the facility account.

OPPORTUNITY ASSESSMENT

Business

Opportunity (Idea) # **44** *Date:* *Your Name:*

Opportunity (Idea) Title: Condense stripper steam and recover benzene

Net Present Value (NPV) for 10 Year Life

$$NPV = a * I - b * (CS + R) - c * WC + d * OTC$$

I = investment: CS = Cost Savings: R = Revenues: WC = Working Capital:
OTC = One Time Costs

Investment Amount		$\leq$ $5 million	$5 to $20 million	$\geq$ $20 million
Discount Rate 10%	a	-0.93	-0.99	-1.0
	b	-4.5	-4.3	-3.8
	c	-0.59	-0.56	-0.49
	d	-0.6	-0.6	-0.56
Discount Rate 25%	a	-0.80	-0.72	-0.67
	b	-2.3	-1.9	-1.4
	c	-0.67	-0.55	-0.41
	d	-0.52	-0.52	-0.44
Discount Rate 40%	a	-0.71	-0.56	-0.49
	b	-1.4	-1.0	-0.66
	c	-0.57	-0.42	-0.27
	d	-0.47	-0.47	-0.35

Discount Rate 10%

$$NPV_{10\%} = \underset{(a)\quad(I)}{0.93 * 50} - \underset{(b)\quad(CS+R)}{4.5 * 108} - \underset{(c)\quad(WC)}{\_\_ * \_\_} + \underset{(d)\quad(OTC)}{\_\_ * \_\_}$$

$$NPV_{10\%} = \underline{\quad 440 \quad} (\$ 1{,}000)$$

Discount Rate 25%

$$NPV_{5\%} = \underset{(a)\quad(I)}{-0.80 * 50} - \underset{(b)\quad(CS+R)}{-2.3 * 108} - \underset{(c)\quad(WC)}{\_\_ * \_\_} + \underset{(d)\quad(OTC)}{\_\_ * \_\_}$$

$$NPV_{25\%} = \underline{\quad 210 \quad} (\$ 1{,}000)$$

Discount Rate 40%

$$NPV_{40\%} = \underset{(a)\quad(I)}{-0.71 * 50} - \underset{(b)\quad(CS+R)}{-1.4 * 108} - \underset{(c)\quad(WC)}{\_\_ * \_\_} + \underset{(d)\quad(OTC)}{\_\_ * \_\_}$$

$$NPV_{40\%} = \underline{\quad 115 \quad} (\$ 1{,}000)$$

OPPORTUNITY ASSESSMENT

Business

Opportunity (Idea) # 22   *Date:*   *Your Name:*

Opportunity (Idea) Title: Fluidized Bed Reactor

Opportunity Assessment Form— Fluidized Bed Reactor

Current Operation:
The reactor has a ceramic fixed-bed catalyst which requires a large volume of air to act as a heat sink to prevent hot spots.

Proposed Operation:
Replace the fixed-bed with a fluidized-bed catalyst.

Assumptions:
The ceramic catalyst can withstand mechanical abrasion. The reaction yield and selectivity are independent of oxygen to inert ratio. The reaction will work at higher reactant concentrations. The fluidized bed requires 70% less gas thus the investment is lower. Assume the investment for the old reactor and new reactor are the same. The compressor requires 70% less power.

Advantages:
The fluidized bed requires much less gas to provide a heat sink for the exothermic reaction. A fluidized bed has a uniform temperature which should result in lower by-product formation, that is decomposition of PROD and REAC at the higher temperatures. Improved temperature control should reduce any tar formation. Reduces end-of-pipe investment by $800,000 and operating cost by $400,000. If the plant is relocated to a new site, the investment for a large fixed bed reactor and a smaller fluidized bed reactor are the same.

Concerns:
If the fixed bed is replaced at the existing site, the investment is $900,000. Possible mechanical abrasion of the catalyst, thus higher operating costs. Need to install a fines collection system, such as a cyclone or bag filter.

National Center for Cleaner Production

OPPORTUNITY ASSESSMENT

Business

Opportunity (Idea) # 22   *Date:*   *Your Name:*

Opportunity (Idea) Title: Fluidized Bed Reactor

Technical Viability: None__Very Low__Low___Medium___High_**X**_Very High___

Implementation: Procedural___Engineering Study_**X**_Research&Development___
Procedural—do in 6 months; Engineering Study—do in 12 months;
Research&Development—do in 2 to 3 years

Benefits of Implementation:

Change in Investment, Operating Costs, Revenues, and Net Present Value with Implementation of the Proposed Operation ($ 1,000)	
Net New Installed Investment	$ 100
Net Reduction in Operating Costs	$ 84 /yr.
Net New Revenues	$ /yr.
Reduction in Working Capital	$
One Time Costs	$
Net Present Value (10 years, 10% Discount)	$ 285
Net Present Value (10 years, 25% Discount)	$ 110
Net Present Value (10 years, 40% Discount)	$ 45

Change to Inputs (Process Feeds, Solvents, Catalysts, etc.) and Outputs (Water wastes, vapor or gas wastes, solid wastes, product(s))	
Inputs:	**Outputs:**
Process Feeds ___ Ton/yr.	Product(s) ___ Ton/yr.
Solvents ___ Ton/yr.	Wastes ___ Ton/yr.
Other ___ Ton/yr.	(Water___ Gas___ Solid ___) [Check appropriate media]

147

OPPORTUNITY ASSESSMENT

Business _____

Opportunity (Idea) # 22   *Date:* _____   *Your Name:* _____

Opportunity (Idea) Title: Fluidized Bed Reactor _____

CALCULATION WORKSHEET

Investment In $1,000   = Estimate * Factor

Use **negative investment for return of an external sale**.

Factors to determine project level Investment	Factor	Type (Check type of estimate input below)
Bare equipment factor	8.0	_____
Skid mounted equipment factor	4.0	_____
Vendor installed factor	2.0	
Project level factor	1.0	X̲

Investment Estimate in $1,000   =   100   Project Level   100 _____

(Estimate * Factor)

Operating Costs   Use dollars per year for the following

Positive numbers are decreased costs, negative numbers are increased costs.

Materials	Quantity Saved/yr.	Units	$/Unit	$ Saved/yr. ($1,000)
Catalyst				-450
Filter bags				-50
Lower thermal oxidizer costs				400

Total material cost Saved   =   -100   ($1,000/yr.)

OPPORTUNITY ASSESSMENT
Business _____

Opportunity (Idea) # **22**   *Date:* _____   *Your Name:* _____

Opportunity (Idea) Title: Fluidized Bed Reactor _____

CALCULATION WORKSHEET (cont.)

Manpower Change—Only include impact if headcount is added or reduced, or if overtime costs are increased decreased   **(negative is an increase)**

	Amount	Factors	$ Saved/yr. ($1,000)
Contractor headcount reduction (Number)		$52,000/yr.	
Employee headcount reduction (Number)		$160,000/yr.	
Contractor overtime reduction (hours/year)		$35 /hr.	
Employee overtime reduction (hours/year)		$35 /hr.	

Total labor cost Saved   = _____ ($1,000/yr)

• •

Positive is a reduction, negative is an increase

Utilities	Amount	Units	$/Unit	$ Saved/yr. ($1,000)
> 400 psig steam		1000# /yr.	3.75	
235 psig steam		1000# /yr.	2.93	
30 psig steam		1000# /yr.	1.84	
Electricity	6132000	KWH/yr.	0.03	194
Air		MSCF/yr.	0.11	
Nitrogen		MSCF/yr.	0.39	

Total Utility Cost Saved   = 184   ($1,000/yr.)

• ▪▪▪▪▪▪▪▪▪▪▪▪ • • • • • • • • • • • •

Total Operating Costs Saved	($1,000/yr)
Materials	$ -100
Manpower	$ _____
Utilities	$ 184
Total	$ 84

OPPORTUNITY ASSESSMENT
Business

Opportunity (Idea) # 22   *Date:*   *Your Name:*

Opportunity (Idea) Title: Fluidized Bed Reactor

CALCULATION WORKSHEET (cont.)

Revenues

Positive is increased revenue, negative is decreased revenue.

Source of Revenue	Amount	Units	$/Unit	$ Earned/yr. ($1,000)
Total Earned =		($1,000/yr.)		

Working Capital Change—Positive is a decreased cost, negative is an increased cost.

Material	Amount Change	Units	$/Unit	Saved / yr. ($1,000)
Total Working Capital Saved	=	($1,000)		

One Time Cost ($ 1,000) $ _____

Positive is increased cost, negative is decreased cost

Enter the one time costs needed for Research and Development, Pilot Plant, Engineering Study, etc. Do not include project team time costs unless these costs are recharged to the facility account.

National Center for Cleaner Production

OPPORTUNITY ASSESSMENT

Business _____

Opportunity (Idea) # __22__   *Date:* _____   *Your Name:* _____

Opportunity (Idea) Title: Fluidized Bed Reactor

Net Present Value (NPV) for 10 Year Life

$$NPV = a * I - b * (CS + R) - c * WC + d * OTC$$

I = investment: CS = Cost Savings: R = Revenues: WC = Working Capital:
OTC = One Time Costs

Investment Amount		$\leq$ $5 million	$5 to $20 million	$\geq$ $20 million
Discount Rate 10%	a	-0.93	-0.99	-1.0
	b	-4.5	-4.3	-3.8
	c	-0.59	-0.56	-0.49
	d	-0.6	-0.6	-0.56
Discount Rate 25%	a	-0.80	-0.72	-0.67
	b	-2.3	-1.9	-1.4
	c	-0.67	-0.55	-0.41
	d	-0.52	-0.52	-0.44
Discount Rate 40%	a	-0.71	-0.56	-0.49
	b	-1.4	-1.0	-0.66
	c	-0.57	-0.42	-0.27
	d	-0.47	-0.47	-0.35

Discount Rate 10%

$$NPV_{10\%} = \underset{(a)\quad(I)}{\underline{0.93*\ 100}} - \underset{(b)\ (CS+R)}{\underline{-4.5\ *\ 84}} - \underset{(c)\ (WC)}{\underline{\quad*\quad}} + \underset{(d)\ (OTC)}{\underline{\quad*\quad}}$$

$NPV_{10\%} =$ __285__   ($ 1,000)

Discount Rate 25%

$$NPV_{5\%} = \underset{(a)\quad(I)}{\underline{-0.80*\ 100}} - \underset{(b)\ (CS+R)}{\underline{-2.3\ *\ 84}} - \underset{(c)\ (WC)}{\underline{\quad*\quad}} + \underset{(d)\ (OTC)}{\underline{\quad*\quad}}$$

$NPV_{25\%} =$ __110__   ($ 1,000)

Discount Rate 40%

$$NPV_{40\%} = \underset{(a)\quad(I)}{\underline{-0.71*\ 100}} - \underset{(b)\ (CS+R)}{\underline{-1.4\ *\ 84}} - \underset{(c)\ (WC)}{\underline{\quad*\quad}} + \underset{(d)\ (OTC)}{\underline{\quad*\quad}}$$

$NPV_{40\%} =$ __45__   ($ 1,000)

OPPORTUNITY ASSESSMENT

Business _____

Opportunity (Idea) # **62**   Date: _____   Your Name: _____

Opportunity (Idea) Title: Install Solvent Contact Scrubber Using REACtant as the Fluid

Opportunity Assessment Form— Install Solvent Contact Scrubber Using REACtant as the Fluid

Current Operation:
Water and acidic water scrubbers are used to cool the gas and remove the solid PROD and liquefied REACtant from the gas stream. A large water stream is produced that has to be steam stripped and sent to wastewater treatment.

Proposed Operation:
Using a high-boiling solvent, the product and reactants can be scrubbed from the gas stream. This would eliminate the water/benzene system. The product stripper would be replaced with an extraction system. In fact, the ideal solvent would be the reactant, because REACtant boils at 120°C and freezes at -30°C.

Assumptions:
Low vapor pressure of the REACtant at –25 °C.
The non-contact condenser is installed in front of this scrubber to reduce the gas temperature from 330 °C to 60 °C.

Advantages:
Remove the extraction system and benzene from the system. Investment savings of $2.5 million for water pretreatment to remove nitrogen compounds from wastewater and for the steam stripper plus thermal oxidizer and reduced operating costs of $200,000. Reduces wastewater treatment from 35 gpm to 1 gpm and reduces the nitrogen content to wastewater treatment by 97%.

Concerns:
Investment of W1.68 billion required for scrubber plus 100 ton (12,000 btu/hr) brine chiller and $500,000 for condenser. Plugging of solvent scrubber with ice or solid PROD.

National Center for Cleaner Production

OPPORTUNITY ASSESSMENT

Business

Opportunity (Idea) # **62**   *Date:*   *Your Name:*

Opportunity (Idea) Title: Install Solvent Contact Scrubber Using REACtant as the Fluid

Technical Viability: None__Very Low__Low___Medium___High_**X**_Very High___

Implementation: Procedural___Engineering Study_**X**_Research&Development___
Procedural—do in 6 months; Engineering Study—do in 12 months;
Research&Development—do in 2 to 3 years

Benefits of Implementation:

Change in Investment, Operating Costs, Revenues, and Net Present Value with Implementation of the Proposed Operation ($ 1,000)	
Net New Installed Investment	$ -600
Net Reduction in Operating Costs	$ 217 /yr.
Net New Revenues	$ /yr.
Reduction in Working Capital	$ 5
One Time Costs	$
Net Present Value (10 years, 10% Discount)	$ 1540
Net Present Value (10 years, 25% Discount)	$ 990
Net Present Value (10 years, 40% Discount)	$ 730

Change to Inputs (Process Feeds, Solvents, Catalysts, etc.) and Outputs (Water wastes, vapor or gas wastes, solid wastes, product(s))

Inputs:		Outputs:		
Process Feeds	___ Ton/yr.	Product(s)	___	Ton/yr.
Solvents	220 Ton/yr.	Wastes	78800	Ton/yr.
Other	___ Ton/yr.	(Water_X_Gas X___Solid ___) [Check appropriate media]		

153

OPPORTUNITY ASSESSMENT

Business

Opportunity (Idea) # 62   *Date:*   *Your Name:*

Opportunity (Idea) Title: Install Solvent Contact Scrubber Using REACtant as the Fluid

CALCULATION WORKSHEET

Investment In $1,000   = Estimate * Factor
Use **negative investment for return of an external sale**.
Factors to determine project level Investment

	Factor	Type (Check type of estimate input below)
Bare equipment factor	8.0	____
Skid mounted equipment factor	4.0	____
Vendor installed factor	2.0	____
Project level factor	1.0	X

Investment Estimate in $1,000   =   1900   Project Level   1900

(Estimate * Factor)

Operating Costs   Use dollars per year for the following
Positive numbers are decreased costs, negative numbers are increased costs.

Materials	Quantity Saved/yr.	Units	$/Unit	$ Saved/yr. ($1,000)
Benzene	438000	lbs.	0.13	57

Total material cost Saved   =   57   ($1,000/yr.)

OPPORTUNITY ASSESSMENT
Business _____

Opportunity (Idea) # 62   Date: _____   Your Name: _____

Opportunity (Idea) Title: *Install Solvent Contact Scrubber Using REACtant as the Fluid*

CALCULATION WORKSHEET (cont.)

Manpower Change—Only include impact if headcount is added or reduced, or if overtime costs are increased decreased   **(negative is an increase)**

	Amount	Factors	$ Saved/yr. ($1,000)
Contractor headcount reduction (Number)	$52,000/yr.		
Employee headcount reduction (Number)	$160,000/yr.		
Contractor overtime reduction (hours/year)	$35 /hr.		
Employee overtime reduction (hours/year)	$35 /hr.		

Total labor cost Saved   = _____ ($1,000/yr)

....

Positive is a reduction, negative is an increase

Utilities	Amount	Units	$/Unit	$ Saved/yr. ($1,000)
> 400 psig steam		1000# /yr.	3.75	
235 psig steam		1000# /yr.	2.93	
30 psig steam		1000# /yr.	1.84	
Electricity	1305250	KWH/yr.	0.03	-40
Air		MSCF/yr.	0.11	
Nitrogen		MSCF/yr.	0.39	

Total Utility Cost Saved   = -40   ($1,000/yr.)

....

Total Operating Costs Saved	($1,000/yr)
Materials	$ 57
Manpower	$
Utilities	$ -40
Total	$ 17

OPPORTUNITY ASSESSMENT
Business _____

Opportunity (Idea) # 62   Date: _____   Your Name: _____

Opportunity (Idea) Title: Install Solvent Contact Scrubber Using REACtant as the Fluid

CALCULATION WORKSHEET (cont.)

Revenues

Positive is increased revenue, negative is decreased revenue.

Source of Revenue	Amount	Units	$/Unit	$ Earned/yr. ($1,000)
_____	_____	_____	_____	_____
_____	_____	_____	_____	_____
_____	_____	_____	_____	_____
_____	_____	_____	_____	_____
Total Earned =	_____	($1,000/yr.)		

Working Capital Change—**Positive is a decreased cost,** negative is an increased cost.

Material	Amount Change	Units	$/Unit	Saved / yr. ($1,000)
Benzene storage	36000	lbs.	0.13	5
_____	_____	_____	_____	_____
_____	_____	_____	_____	_____
Total Working Capital Saved		= 5	($1,000)	

One Time Cost ($ 1,000) $ _____

Positive is increased cost, negative is decreased cost

Enter the one time costs needed for Research and Development, Pilot Plant, Engineering Study, etc. Do not include project team time costs unless these costs are recharged to the facility account.

OPPORTUNITY ASSESSMENT
Business

Opportunity (Idea) # 62   Date: Your Name:

Opportunity (Idea) Title: *Install Solvent Contact Scrubber Using REACtant as the Fluid*

Net Present Value (NPV) for 10 Year Life

$$NPV = a * I - b * (CS + R) - c * WC + d * OTC$$

I = investment: **CS** = Cost Savings: **R** = Revenues: **WC** = Working Capital:
OTC = One Time Costs

Investment Amount		$\leq$ \$5 million	\$5 to \$20 million	$\geq$ \$20 million
Discount Rate 10%	*a*	-0.93	-0.99	-1.0
	b	-4.5	-4.3	-3.8
	c	-0.59	-0.56	-0.49
	d	-0.6	-0.6	-0.56
Discount Rate 25%	*a*	-0.80	-0.72	-0.67
	b	-2.3	-1.9	-1.4
	c	-0.67	-0.55	-0.41
	d	-0.52	-0.52	-0.44
Discount Rate 40%	*a*	-0.71	-0.56	-0.49
	b	-1.4	-1.0	-0.66
	c	-0.57	-0.42	-0.27
	d	-0.47	-0.47	-0.35

Discount Rate 10%
$NPV_{10\%}$ = $\underline{\text{0.93* -600}}$ - $\underline{\text{4.5 * 217}}$ - $\underline{\text{-0.59 * 5}}$ + $\underline{\text{*}}$
      (*a*) (**I**)   (*b*) (**CS** + **R**)   (*c*) (**WC**) (*d*) (**OTC**)

$NPV_{10\%}$ =**1540**   (\$ 1,000)

Discount Rate 25%
$NPV_{5\%}$ = $\underline{\text{-0.80 *-600}}$ - $\underline{\text{-2.3 * 217}}$ - $\underline{\text{-0.67 * 5}}$ + $\underline{\text{*}}$
      (*a*) (**I**)   (*b*) (**CS** + **R**)   (*c*) (**WC**) (*d*) (**OTC**)

$NPV_{25\%}$ =___ **990**   (\$ 1,000)

Discount Rate 40%
$NPV_{40\%}$ = $\underline{\text{-0.71 *-600}}$ - $\underline{\text{-1.4 * 217}}$ - $\underline{\text{-0.57 * 5}}$ + $\underline{\text{*}}$
      (*a*) (**I**)   (*b*) (**CS** + **R**)   (*c*) (**WC**) (*d*) (**OTC**)

$NPV_{40\%}$ = **730**   (\$ 1,000)

OPPORTUNITY ASSESSMENT

Business

Opportunity (Idea) # 23   *Date:*   *Your Name:*

Opportunity (Idea) Title: Install Non-contact Condenser in Line to Scrubber

Opportunity Assessment Form—Install Non-contact Condenser in Line to Scrubber

Current Operation:
The reactor hot exit gas is first cooled by a process/process heat exchanger. The cooler gas (about 300 °C) is then cooled by the water scrubber.

Proposed Operation:
The PRODuct freezes at about 60°C. A cooler operating at about 65°C would condense the PRODuct with little condensation of the REACtant.

Assumptions:
Low solubility of noxious gases in the liquid.

Advantages:
Eliminate loss of product leaving with the water from the benzene extractor.

Concerns:
Possible formation of higher molecular weight solids.

National Center for Cleaner Production

OPPORTUNITY ASSESSMENT
Business

Opportunity (Idea) # **23**   *Date:*   *Your Name:*

Opportunity (Idea) Title: Install Non-contact Condenser in Line to Scrubber

Technical Viability: None__Very Low__Low___Medium___High__Very High_**X**_

Implementation: Procedural___Engineering Study_**X**_Research&Development___
Procedural—do in 6 months; Engineering Study—do in 12 months;
Research&Development—do in 2 to 3 years

Benefits of Implementation:

Change in Investment, Operating Costs, Revenues, and Net Present Value with Implementation of the Proposed Operation ($ 1,000)	
Net New Installed Investment	$ 500
Net Reduction in Operating Costs	$ -30 /yr.
Net New Revenues	$ 48 /yr.
Reduction in Working Capital	$
One Time Costs	$
Net Present Value (10 years, 10% Discount)	$ -380
Net Present Value (10 years, 25% Discount)	$ -360
Net Present Value (10 years, 40% Discount)	$ -330

Change to Inputs (Process Feeds, Solvents, Catalysts, etc.) and Outputs (Water wastes, vapor or gas wastes, solid wastes, product(s))			
Inputs:		**Outputs:**	
Process Feeds ___ Ton/yr.		Product(s) 4.8 Ton/yr.	
Solvents ___ Ton/yr.		Wastes ___ Ton/yr.	
Other ___ Ton/yr.		(Water__Gas __Solid __) [Check appropriate media]	

OPPORTUNITY ASSESSMENT
Business _____

Opportunity (Idea) # 23 _____ Date: _____ Your Name: _____

Opportunity (Idea) Title: *Install Non-contact Condenser in Line to Scrubber* _____

CALCULATION WORKSHEET

Investment In $1,000   = Estimate * Factor
Use **negative investment for return of an external sale**.

Factors to determine project level Investment

	Factor	Type (Check type of estimate input below)
Bare equipment factor	8.0	____
Skid mounted equipment factor	4.0	____
Vendor installed factor	2.0	____
Project level factor	1.0	X

Investment Estimate in $1,000   =   500   Project Level   500 _____

(Estimate * Factor)

Operating Costs   Use dollars per year for the following
Positive numbers are decreased costs, negative numbers are increased costs.

Materials	Quantity Saved/yr.	Units	$/Unit	$ Saved/yr. ($1,000)
_____	_____	____	____	_____
_____	_____	____	____	_____
_____	_____	____	____	_____
_____	_____	____	____	_____
_____	_____	____	____	_____
_____	_____	____	____	_____
_____	_____	____	____	_____

Total material cost Saved   =   57   ($1,000/yr.)

OPPORTUNITY ASSESSMENT
Business

Opportunity (Idea) # 23   Date: Your Name:

Opportunity (Idea) Title: *Install Non-contact Condenser in Line to Scrubber*

CALCULATION WORKSHEET (cont.)

Manpower Change—Only include impact if headcount is added or reduced, or if overtime costs are increased decreased **(negative is an increase)**

	Amount	Factors$ Saved/yr. ($1,000)
Contractor headcount reduction (Number)		$52,000/yr.
Employee headcount reduction (Number)		$160,000/yr.
Contractor overtime reduction (hours/year)		$35 /hr.
Employee overtime reduction (hours/year)		$35 /hr.

Total labor cost Saved   =_____ **($1,000/yr)**

••

Positive is a reduction, negative is an increase

Utilities	Amount	Units	$/Unit	$ Saved/yr. ($1,000)
> 400 psig steam		1000# /yr.	3.75	
235 psig steam		1000# /yr.	2.93	
30 psig steam		1000# /yr.	1.84	
Electricity		KWH/yr.	0.03	-30
Air		MSCF/yr.	0.11	
Nitrogen		MSCF/yr.	0.39	

Total Utility Cost Saved   = **-30**   **($1,000/yr.)**

••

Total Operating Costs Saved		($1,000/yr)
Materials	$ _____	
Manpower	$ _____	
Utilities	$ **-30**_____	
Total	$ **-30**_____	

National Center for Cleaner Production

OPPORTUNITY ASSESSMENT
Business

Opportunity (Idea) #   23   *Date:*   *Your Name:*

Opportunity (Idea) Title: Install Non-contact Condenser in Line to Scrubber

CALCULATION WORKSHEET (cont.)

Revenues
Positive is increased revenue, negative is decreased revenue.

Source of Revenue	Amount	Units	$/Unit	$ Earned/yr. ($1,000)
Recovered product	1.1	lbs./hr.	$5.00/lb.	48

Total Earned   =   48   ($1,000/yr.)

Working Capital Change—Positive is a decreased cost, negative is an increased cost.

Material	Amount Change	Units	$/Unit	Saved / yr. ($1,000)

Total Working Capital Saved   =   ($1,000)

One Time Cost ($ 1,000) $
Positive is increased cost, negative is decreased cost
Enter the one time costs needed for Research and Development, Pilot Plant, Engineering Study, etc. Do not include project team time costs unless these costs are recharged to the facility account.

National Center for Cleaner Production

OPPORTUNITY ASSESSMENT

Business _____

Opportunity (Idea) # **23**   Date: _____   Your Name: _____

Opportunity (Idea) Title: **Install Non-contact Condenser in Line to Scrubber** _____

Net Present Value (NPV) for 10 Year Life

$$NPV = a * I - b * (CS + R) - c * WC + d * OTC$$

I = investment: **CS** = Cost Savings: **R** = Revenues: **WC** = Working Capital:
OTC = One Time Costs

Investment Amount		$\leq$ $5 million	$5 to $20 million	$\geq$ $20 million
Discount Rate 10%	a	-0.93	-0.99	-1.0
	b	-4.5	-4.3	-3.8
	c	-0.59	-0.56	-0.49
	d	-0.6	-0.6	-0.56
Discount Rate 25%	a	-0.80	-0.72	-0.67
	b	-2.3	-1.9	-1.4
	c	-0.67	-0.55	-0.41
	d	-0.52	-0.52	-0.44
Discount Rate 40%	a	-0.71	-0.56	-0.49
	b	-1.4	-1.0	-0.66
	c	-0.57	-0.42	-0.27
	d	-0.47	-0.47	-0.35

Discount Rate 10%
$NPV_{10\%}$ = <u>0.93* 500</u> - <u>-4.5</u> * <u>-30+48-</u> ____ * ____ + ____ * ____
            (a)   (I)   (b)   (CS + R)   (c)   (WC)   (d)   (OTC)

$NPV_{10\%}$ =<u>-380</u>   ($ 1,000)

Discount Rate 25%
$NPV_{5\%}$ = <u>-0.80 *500</u> - <u>-2.3</u> * <u>-30+48 -</u> ____ * ____ + ____ * ____
            (a)   (I)   (b)   (CS + R)   (c)   (WC)   (d)   (OTC)

$NPV_{25\%}$ = <u>-360</u>   ($ 1,000)

Discount Rate 40%
$NPV_{40\%}$ = <u>-0.71 *500</u> - <u>-1.4</u> * <u>-30+48-</u> ____ * ____ + ____ * ____
             (a)   (I)   (b)   (CS + R)   (c)   (WC)   (d)   (OTC)

$NPV_{40\%}$ = <u>-330</u>   ($ 1,000)

All Ideas Proposed

About 80 to 100 ideas were generated in the brainstorming session. The following list contains those ideas that were considered technically feasible.

Acid Scrubber Off-Gas Stream (A10)

- Use pure oxygen in the reactor instead of air to eliminate N_2.
- Recycle the acid scrubber off-gas (stream A10) to the reactor and add pure O_2 as make-up.
- Use a new or improved catalyst in the reactor.
- Replace the existing fixed bed reactor with a fluidized-bed reactor to eliminate hot spots that lead to unwanted byproduct formation.
- Add a second heat exchanger in series with the preheater to improve energy recovery.
- Use a different heat sink instead of nitrogen, such as CO_2 or steam.
- Use an indirect-contact heat exchanger to cool the reactor off-gas and condense the product and reactant.

Wastewater Stream from Benzene Extractor (A22)

- Combine the acid and water scrubbers to reduce water consumption.
- Recycle the water stream from the benzene extractor (A22) to the water scrubber.
- Recycle the water stream from the steam stripper (A41) to the water scrubber.
- Use caustic soda instead of ammonia for pH control of the water stream to the benzene extractor (stream A6) to reduce total nitrogen load to wastewater treatment.
- Replace benzene with a different extraction solvent, such as toluene or xylene.
- Freeze-crystallize REAC and PROD from the water to improve product recovery and reduce total nitrogen load to wastewater treatment.
- Use a multi-effect evaporator to concentrate stream A41.
- Use the chilled reactant (REAC) to scrub itself from the reactor off-gas.
- Use a condenser and decanter to separate and recycle the benzene from the steam stripper overheads instead of a thermal oxidizer.

Information Package Sent to Brainstorming Team

Waste Minimization Program Purpose

A chemical intermediate is manufactured by a 40-year-old process that generates an unacceptable level of waste compared to current standards of environmental performance. The facility is located within city limits, and new local environmental regulations require that an expensive end-of-pipe treatment device (requiring investment of $1.6 million with a $560,000 / yr. operating cost) be installed on a gaseous emission source unless more economical source reduction measures can be identified and implemented. The local community is unaware that the process emits noxious compounds—thus, if the business is unable to convince the community that it will adequately treat the noxious compounds, then the business may need to relocate the manufacturing process to a new site.

Also, the nitrogen content of the wastewater stream from the process was too high. New wastewater regulations would require lower nitrogen content in the outfall from the wastewater treatment system, and this process is the largest contributor to the nitrogen content in the wastewater. The wastewater from the process would require pretreatment before being sent to the treatment system at an investment of $400,000 and a $50,000 / yr operating cost.

The separation steps involved in the existing process include: (1) aqueous scrubbing of a reactor off-gas to condense and capture the reactant and product, (2) extraction of the reactant and product into a benzene solvent, and (3) high-purity distillation of the reactant and product. Benzene is recovered for recycle, and the aqueous raffinate stream is stripped of benzene before discharge to a biological wastewater treatment plant.

The need to reduce the high investment associated with end-of-pipe treatment of gaseous and wastewater emissions provide a great opportunity to critically review the process for optimal design. The review will target process changes that can achieve significant environmental and financial gains. Potential benefits include reduced capital investment, lower operating cost, waste reduction, process simplification, and elimination of benzene handling.

Thus, the purpose of this review is to identify process improvements that will:

- Reduce the end-of-pipe investment requirements to treat a toxic gas stream, and

- Reduce the nitrogen content in the outfall from the wastewater treatment facility.

Brainstorming Session Purpose and Products

PURPOSE

To identify ideas that reduce toxic gaseous and nitrogen bearing water wastes.

WE WOULD LIKE TO:

- Take advantage of your perspective and expertise.

- Identify all ideas that reduce the amount of waste generated.

- Use the synergy of a brainstorming session to identify cost-effective ideas.

- Address any barriers and concerns.

BENEFITS

The business can realize the desired emissions and energy reductions and improve the operation and utility of the process with maximum return to the stockholders.

PRODUCTS

- Identification of the changes (technical and operational) required to improve operation of the process.

- A prioritized list of opportunities and recommendations to be considered by the business and manufacturing site to reduce waste generation.

- A path forward with responsibilities assigned.

Brainstorming Session Agenda

Day 1 —Date

Plant Tour		11:00 a.m.
Introductions	All	1:00 p.m.
Review Waste Minimization Methodology	Facilitator	1:15 p.m.
Review Agenda, Purpose and Products and Ground Rules	Facilitator	2:00 p.m.
Break		2:15 p.m.
Business and Environmental Drivers	Business Leader	2:30 p.m.
Process Overview and Sources of Waste	Process Expert	2:50 p.m.

Day 2 —Date

Process Overview	Process Expert	8:00 a.m.
Idea Generation	Facilitator	8:15 a.m.
Break		10:00 a.m.
Idea Generation		10:15 a.m.
Lunch		12:00 p.m.
Idea Generation		12:30 p.m.
Break		2:00 p.m.
Develop Ranking Criteria		2:15 p.m.
Rank Ideas	All	2:30 p.m.
Close		4:00 p.m.

Day 3 —Date

Idea Development and Assessment Overview	Facilitator	8:00 a.m.
Idea Development and Assessment	All	8:30 a.m.
Review of Assessments	Facilitator	2:00 p.m.
Close		2:30 p.m.

Participants Responsibilities

To maximize the contribution of the group, each individual group member needs to follow three ground rules—participate, be concise, and be additive.

First, each individual invited to the session is expected to participate—silence is unacceptable. All participants need to understand that their ideas, no matter how off-the-wall they may sound at first, will not be judged during the brainstorming portion of the meeting. The judgment and critique will be reserved for the screening portion of the session. The concept behind brainstorming is that one idea should lead to a new idea or build on the previous one. If a person does not speak up, then that individual's ingenuity is not being fully exercised.

Second, participants need to be concise. Ideas must be conveyed clearly and completely. Answer any questions on the meaning of your idea, but do not engineer the idea. There just is not enough time during the brainstorming portion of the meeting to engineer every idea. In addition, it will restrict the flow of new ideas.

Third, participants should be additive and avoid critiquing other people's ideas. Sometimes, an idea that was tried in the past and failed for either technological or political reasons will work in the current climate. Also keep in mind that all ideas will be reexamined at a later date. The goal of the brainstorming session is to get all possible ideas on the table, so that the best idea can be evaluated and chosen during the screening and evaluation stages of the assessment phase.

Problem Definition

To be effective in a brainstorming session, the participants, that is **you**, must study the information package before coming to the session. Because 100 to 200 ideas will be generated in a short time period, an unprepared participant will contribute less than a well-prepared participant. One proven approach to aid in your participation is to use the waste stream and process analysis techniques given below. The purpose of these analyses is for you to ask yourself the right questions. With the right questions the best waste minimization solutions will almost always become obvious.

There are two general categories of root causes for process waste—operational and fundamental. Operational root causes arise from how the process is operating versus how it should operate. Examples of operational root causes are:

- Poor understanding of the process,

- Control problems,

- Not following operating procedures, and

- Maintenance problems.

Fundamental root causes arise from the chemistry and thermodynamic limitations of the process. Examples of fundamental root causes are:

- Chemistry route picked that requires toxic solvents,

- Catalyst selection and byproduct formation,

- Reactor operation and byproduct formation, and

- Not understanding the functions and thermodynamic principles of the separation processes.

Note—Basic thermodynamic principles state that any material introduced to or created in a process will escape as a waste or will not be completely destroyed by an end-of-pipe treatment device.

As part of this information packet for the opportunity identification team, the facilitator, champion and process expert have developed a series of question for the participants to consider. These questions will help trigger your creative thought processes.

List of Initial Questions

Why use air as the source of oxygen?

If the nitrogen is required as a heat sink for the exothermic reaction, what other means of removing the heat of reaction are available?

Is there another heat sink that could be used?

Why use water as the cooling and scrubbing medium?

Why use fresh water?

Why a separate acid scrubber?

How can the reactor temperature control be improved?

Why use benzene?

How does one reduce the nitrogen content in the water stream going to wastewater treatment?

Waste Stream Analyses

The best pollution prevention options cannot be implemented unless they are identified. To uncover the best options, one approach is to focus on the waste streams themselves, using the following four steps:

1) In Figures B-1 and B-2 the components are listed for streams A10 (Acid Scrubber Vent) and A22 (Benzene Extractor Bottoms).

2) Identify the compounds triggering the concern and determine the sources of these compounds within the process. The compounds of concern are for

- Stream A10—HCN and the nitrogen- and sulfur-containing compounds from feeds, reactor and acid scrubber.

- Stream A22—Benzene and nitrogen-bearing compounds from extractor, feeds and reactor.

3) Identify the highest volume materials. The sources of these diluents within the process are for

- Stream A10—Nitrogen and unreacted oxygen from air fed to the compressor.

- Stream A22—Water flow from the off-gas scrubbers.

Figure B-1 Waste Stream Description

Date: January 15, 1998

Process: Intermediates Manufacture

Waste Stream ID **Before** any Treatment Device: Stream A10 – Acid Scrubber Vent

Waste Description: Gas stream Containing Air and Toxic Compounds

Waste Treatment (yes or no): No

High Toxicity (yes or no): Yes
Special Safety Hazard (yes or no): No

Waste Compounds and Composition in order of Importance

Compound Name List in order from high to low flow rates	Wt%	Listed Yes or No**	Waste Origin that is where it is introduced to the process, or is introduced to the waste stream.
Nitrogen	72%	No	Feed to reactor
Oxygen	22%	No	Feed to reactor
Water	6%	No	Water scrubbers and water of reaction
IMPURity	80 ppmw	No	Feed to reactor
NO2, SO2, COS	50 ppmw	Yes	Reactor byproducts
CO2, HCN	86 ppmw	Yes	Reactor byproducts

** Listed – on any regulatory lists

Stream Total Flow: 51344

Units of Flow: Lbs./hr.

Other items of concern (odor, pH, etc.): High toxicity

Figure B-2 Waste Stream Description

Date: January 15, 1998

Process: Intermediates Manufacture

Waste Stream ID **Before** any Treatment Device: Stream A22 – Benzene Extractor Bottoms

Waste Description: Wastewater stream containing benzene and nitrogen bearing compounds

Waste Treatment (yes or no): yes

High Toxicity (yes or no): yes

Special Safety Hazard (yes or no): no

Waste Compounds and Composition in order of Importance

Compound Name List in order from high to low flow rates	Wt%	Listed Yes or No**	Waste Origin that is where it is introduced to the process, or is introduced to the waste stream.
Water	99%	No	Water scrubbers and water of reaction
Ammonium Sulfate	0.44%	No	Acid scrubber neutralization
Benzene	0.22%	Yes	Benzene extractor
Ammonia	529 ppmw	Yes	pH control, added before benzene extractor
HCN	261 ppmw	Yes	Reactor byproducts
SO2 and NO2 each	154 ppmw	Yes	Reactor byproducts
REAC + PROD	187 ppmw	No	Feed and reactor product

** Listed – on any regulatory lists

Stream Total Flow: 14938

Units of Flow: lbs./hr.

Other items of concern (odor, pH, etc.):

Process Analysis

For either a new or existing process, the following steps are taken:

1) List all raw materials reacting to salable products, any intermediates, and all salable products. This is "List 1."

- For the case study process, the salable compound is PROD, and the raw materials are REAC, oxygen, and ammonia.

2) List all other materials in the process, such as nonsalable byproducts, solvents, water, air, nitrogen, acids, bases, and so on. This is "List 2."

- These other components (water, nitrogen, excess oxygen, reactor byproducts, benzene, etc.) are not salable or do not make a salable product. They are only necessary because of the chemistry and engineering technologies that were chosen in the development of the original product and process.

3) For each compound in List 2, ask "How can I use a material from List 1 to do the same function of the compound in List 2?" or "How can I modify the process to eliminate the need for the material in List 2?"

4) For those materials in List 2 that are the result of producing non-salable products (i.e., waste byproducts), ask, "How can the chemistry or process be modified to minimize or eliminate the wastes (for example, 100% reaction selectivity to a desired product)?"

FORM: Process Constituents and Sources

Process: Intermediates Manufacture   Date: 15 January, 1998

List 1

Constituent:	Source (Feed, Reactor, Unit Operation):

Salable Products

PROD	Reactor

Intermediates (result in salable products)

Essential Feed Materials (only those constituents used to produce the intermediates and salable products)

REAC	Feed
Ammonia	Feed
Oxygen	Air feed to reactor

List 2

Constituent:	Source (Feed, Reactor, Unit Operation):

Other Materials (Non-salable byproducts, solvents, water, air, nitrogen, acids, etc.)

Nitrogen	Air feed to reactor
IMPURity	Feed
Water	Feed to scrubbers
Benzene	Extractor solvent
NO2, SO2, COS, CO2, HCN	Reactor byproducts
Ammonium sulfate	Acid scrubber
Ammonia	Added before extractor

Typical Questions for Each Participant to Consider

Each participant is expected to bring a different perspective to all ideas that are generated. This is why it is so important to pick the right mix of people for the session. The concept is to have everyone be aware of the interdependency of any one idea on the whole, and how that idea can impact other ideas.

For the process engineer: What is the life of the present process? The present product? What is the competition doing that the group should know about?

For the chemist: What are the principal factors affecting yield, conversion, and selectivity? If the reaction is reversible, can byproducts be back-reacted to the incoming feed materials or converted to other useable products? If non-salable products are homologues of the reactants or intermediates, how can they be converted and recycled? What other catalysts are possible? If excess reactants or inerts are being used, ask why? If air, water, or a solvent is being used, ask why?

For the separations specialist: If an exit gas or water stream is being generated, what other separation techniques could be used to eliminate the stream? For trace levels of contaminants, how can the separation unit operations be improved? If large amounts of energy are required, what other separation technologies are applicable? If significant heating, followed by cooling, and then reheating takes place, what other combinations of unit operations can be used to minimize energy usage?

For the environmental specialist: What are the hazardous, carcinogenic, or toxic materials in the waste and product streams that require or could require further treatment? What are the present and future (5-10 years out) environmental laws that impact the waste from this process? What end-of-pipe technologies are appropriate?

For the engineering evaluator: For the current waste streams, what are the end-of-pipe treatment costs? What is the cost of waste generation for the current process?

For the energy specialist: What are the opportunities to save energy in the process? What are the process-to-process energy exchange opportunities? What are the corporate energy goals?

For the lead operator and maintenance representative: What operating procedures are outdated or not followed? How does poor operation affect waste generation? How can startup, shutdown, and maintenance wastes be reduced?

Process Flowsheet

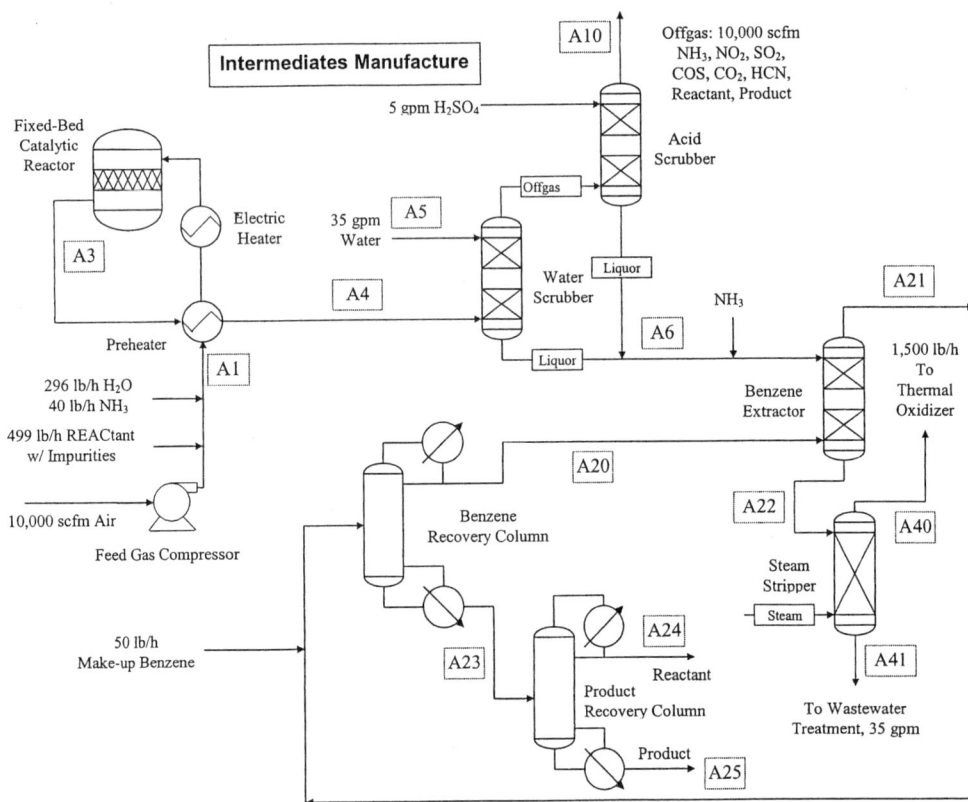

Intermediates Manufacture

Offgas: 10,000 scfm
NH_3, NO_2, SO_2,
COS, CO_2, HCN,
Reactant, Product

A10

5 gpm H_2SO_4

Acid
Scrubber

Fixed-Bed
Catalytic
Reactor

Electric
Heater

35 gpm
Water

A5

A3

Offgas

A4

Water
Scrubber

Liquor

A6

NH₃

A21

Preheater

A1

Liquor

1,500 lb/h
To
Thermal
Oxidizer

296 lb/h H_2O
40 lb/h NH_3

Benzene
Extractor

499 lb/h REACtant
w/ Impurities

A20

A22

A40

10,000 scfm Air

Benzene
Recovery Column

Steam
Stripper

Feed Gas Compressor

Steam

A24

A41

50 lb/h
Make-up Benzene

A23

Reactant

To Wastewater
Treatment, 35 gpm

Product
Recovery Column

Product

A25

Flow Chart

Waste Reduction for Cleaner Production Improvement
Business Unit—Intermediates Manufacture

Intermediates Manufacturing Process Flow Chart

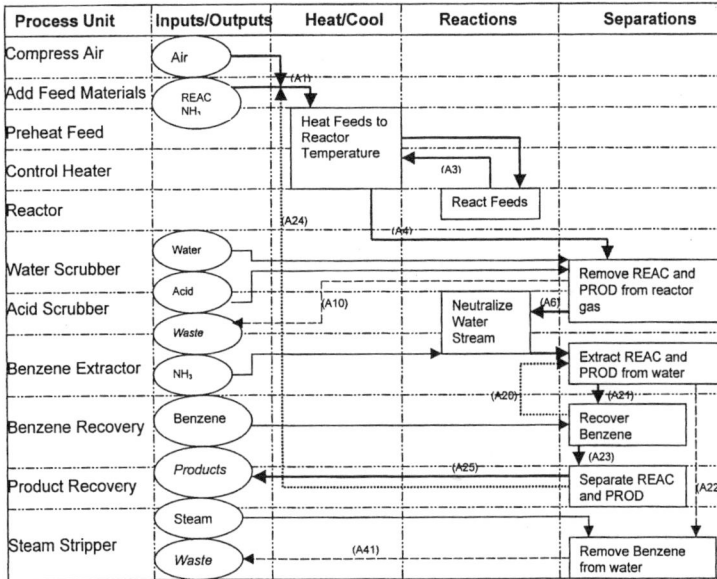

Process Unit	Inputs/Outputs	Heat/Cool	Reactions	Separations
Compress Air	Air			
Add Feed Materials	REAC NH₃	(A1)		
Preheat Feed		Heat Feeds to Reactor Temperature		
Control Heater			(A3)	
Reactor		(A24)	React Feeds	
		(A4)		
Water Scrubber	Water			Remove REAC and PROD from reactor gas
Acid Scrubber	Acid	(A10)	Neutralize Water Stream (A6)	
	Waste			
Benzene Extractor	NH₃			Extract REAC and PROD from water
			(A20)	(A21)
Benzene Recovery	Benzene			Recover Benzene
				(A23)
Product Recovery	Products		(A25)	Separate REAC and PROD (A22)
Steam Stripper	Steam			
	Waste	(A41)		Remove Benzene from water

Process Flowsheet and Flow Chart Function Forms

Function Description Form

Flowsheet Designation: Feed Gas Compressor

Type of Unit (reactor, distillation column, heater, tank, and so on):
Centrifugal compressor

Function of Unit: To compress air to reactor pressure of 75 psia

Principal Control Parameter(s): (1) Maintain above a minimum gas flow rate. (2) Outlet pressure

Principal Poor-operation Problems: Mechanical outage of the compressor

Uptime (Average time between outages): Yearly preventative maintenance

Wastes going to treatment or being emitted to the environment
    Process (total lbs./hr):
        Vapor or Gas:
        Liquid (non-aqueous):
        Water type:
        Solid:
        Hazardous:

Other
    Uptime losses (lbs. Generated at Startup and Shutdown):
    Maintenance losses: 50 lbs. every year
    Storage losses:

Function Description Form

Flowsheet Designation: Pre-heater and Electric Heater

Type of Unit (reactor, distillation column, heater, tank, and so on): Finned tubular heat exchanger and resistance electric heater

Function of Unit: To recover heat from hot reactor exit gas and to control the reactor gas inlet temperature

Principal Control Parameter(s): (1) Cleanliness of heat exchanger (2) Electric heater outlet temperature

Principal Poor-operation Problems: Fouled heat exchanger and burned out heater electrodes

Uptime (Average time between outages): Yearly preventative maintenance

Wastes going to treatment or being emitted to the environment
    Process (total lbs./hr):
        Vapor or Gas:
        Liquid (non-aqueous):
        Water type:
        Solid:
        Hazardous:

Other
    Uptime losses (lbs. Generated at Startup and Shutdown):
    Maintenance losses:
    Storage losses:

Function Description Form

Flowsheet Designation: Fixed Bed Catalytic Reactor

Type of Unit (reactor, distillation column, heater, tank, and so on): Reactor with a fixed bed ceramic catalyst

Function of Unit: To react oxygen, ammonia, REAC to PROD

Principal Control Parameter(s): (1) Gas space velocity, that is, the flow rate of the materials through the bed. (2) Inlet temperature of the reactants to the reactor.

Principal Poor-operation Problems: (1) Deactivation of the catalyst through fouling of the catalyst by tar formation on the catalyst. (2) Waste product formation by hotspots in the bed.

Uptime (Average time between outages): 6 months between burnout of the catalysts. Replace catalyst every two years.

Wastes going to treatment or being emitted to the environment
     Process (total lbs./hr):
          Vapor or Gas:
          Liquid (non-aqueous):
          Water type:
          Solid:
          Hazardous:

Other
     Uptime losses (lbs. Generated at Startup and Shutdown):
     Maintenance losses: 50 lbs. every 6 months. 1,000 lbs. every 2 years
     Storage losses:

Function Description Form

Flowsheet Designation: Water Scrubber

Type of Unit (reactor, distillation column, heater, tank, and so on): Packed bed water scrubber

Function of Unit: To cool the gas stream and remove REAC and PROD from gas stream

Principal Control Parameter(s): Water flow rate

Principal Poor-operation Problems: Fouling of beds

Uptime (Average time between outages): Yearly preventative maintenance

Wastes going to treatment or being emitted to the environment
      Process (total lbs./hr):
            Vapor or Gas:
            Liquid (non-aqueous):
            Water type:
            Solid:
            Hazardous:

Other
      Uptime losses (lbs. Generated at Startup and Shutdown):
      Maintenance losses:
      Storage losses:

Function Description Form

Flowsheet Designation: Acid Scrubber

Type of Unit (reactor, distillation column, heater, tank, and so on): Packed bed acidic water scrubber

Function of Unit: To remove trace amounts of REAC and PROD from gas stream

Principal Control Parameter(s): (1) water pH (2) Water flow rate

Principal Poor-operation Problems: (1) pH control system (2) occasional collapse of the beds

Uptime (Average time between outages): Yearly preventative maintenance

Wastes going to treatment or being emitted to the environment
> **Process (total lbs./hr):** 51344
>> **Vapor or Gas:** 218 lbs./hr of contaminants and toxic compounds
>> **Liquid (non-aqueous):**
>> **Water type:**
>> **Solid:**
>> **Hazardous:**

Other
> **Uptime losses (lbs. Generated at Startup and Shutdown):**
> **Maintenance losses:**
> **Storage losses:**

Function Description Form

Flowsheet Designation: Benzene Extractor

Type of Unit (reactor, distillation column, heater, tank, and so on): Trayed liquid/liquid extractor

Function of Unit: To remove REAC and PROD from water stream

Principal Control Parameter(s): Benzene flow rate

Principal Poor-operation Problems: none

Uptime (Average time between outages): Yearly preventative maintenance

Wastes going to treatment or being emitted to the environment
  Process (total lbs./hr): 14940
      Vapor or Gas: Fugitive emissions
      Liquid (non-aqueous):
      Water type: 14940 lbs./hr to steam stripper to remove 33.4 lbs./hr
          benzene
      Solid:
      Hazardous:

Other
      Uptime losses (lbs. Generated at Startup and Shutdown):
      Maintenance losses:
      Storage losses:

Function Description Form

Flowsheet Designation: Benzene Recovery

Type of Unit (reactor, distillation column, heater, tank, and so on): Trayed distillation column

Function of Unit: To remove benzene from REAC and PROD

Principal Control Parameter(s): Condenser temperature and reflux rate

Principal Poor-operation Problems: Tar buildup in the reboiler

Uptime (Average time between outages): Flush reboiler every 6 months

Wastes going to treatment or being emitted to the environment
      Process (total lbs./hr): 1 to 3
            Vapor or Gas: Condenser vent
            Liquid (non-aqueous):
            Water type:
            Solid:
            Hazardous:

Other
      Uptime losses (lbs. Generated at Startup and Shutdown):
      Maintenance losses: 200 lbs. tar every 6 months
      Storage losses:

Function Description Form

Flowsheet Designation: Product Recovery

Type of Unit (reactor, distillation column, heater, tank, and so on): Trayed distillation column

Function of Unit: To separate REAC and PROD

Principal Control Parameter(s): Condenser temperature and reflux rate

Principal Poor-operation Problems: Tar buildup in the reboiler

Uptime (Average time between outages): Flush reboiler every 6 months

Wastes going to treatment or being emitted to the environment
    Process (total lbs./hr):
        Vapor or Gas:
        Liquid (non-aqueous):
        Water type:
        Solid:
        Hazardous:

Other
    Uptime losses (lbs. Generated at Startup and Shutdown):
    Maintenance losses: 400 lbs. tar every 6 months
    Storage losses:

Function Description Form

Flowsheet Designation: Steam Stripper

Type of Unit (reactor, distillation column, heater, tank, and so on): Packed column

Function of Unit: To remove benzene from the water to wastewater treatment

Principal Control Parameter(s): Exit water temperature

Principal Poor-operation Problems: Occasional high gas rate up column and disruption of the packing

Uptime (Average time between outages): Yearly maintenance

Wastes going to treatment or being emitted to the environment
　　　Process (total lbs./hr): 17948
　　　　　　Vapor or Gas: 1551 lbs./hr steam plus benzene
　　　　　　Liquid (non-aqueous):
　　　　　　Water type: 16400 lbs./hr to wastewater treatment
　　　　　　Solid:
　　　　　　Hazardous:

Other
　　　Uptime losses (lbs. Generated at Startup and Shutdown):
　　　Maintenance losses: 400 lbs. tar every 6 months
　　　Storage losses:

Mass Balances

Intermediates Manufacture Mass Balance

Stream Number	A1	A3	A4	A5	A6	A10	A20	A21	A22	A23
Phase	Vapor	Vapor	Vapor	Liquid	Liquid	Vapor	Liquid	Liquid	Liquid	Liquid
Component					**Mass Flow (lb/hr)**					
Benzene (C_6H_6)	0	0	0	0	0	0	4949.6	4949.6	33.4	0.0089
Nitrogen (N_2)	36952	36952	36952	0	0.22	36952	0.22	0.22	0.0024	0
Oxygen (O_2)	11298	11064	11064	0	0.053	11064	0.053	0.053	0.00018	0
Water (H_2O)	296	433.3	433.3	17500	14821	3112.2	15.9	15.9	14820	0
REACtant	483	205.8	205.8	0	202.8	3	2.7	203.1	1.7	200.4
PRODuct	1	207.7	207.7	0	207.7	0	1.4E-05	206.6	1.1	206.6
IMPURity	15	15	15	0	2.4	12.6	0.52	2.4	0.013	1.9
Ammonia (NH_3)	40	8	8	0	0	0	0.13	0.13	7.9	0
Nitric Oxide (NO_2)	0	14	14	0	2.3	11.7	0.21	0.21	2.1	0
Sulfur Dioxide (SO_2)	0	39.1	39.1	0	2.9	36.2	0.27	0.27	2.6	0
Carbonyl Sulfide (COS)	0	18.3	18.3	0	0.0018	18.3	1.6E-04	1.6E-04	0.0016	0
Carbon Dioxide (CO_2)	0	120.9	120.9	0	0.025	120.9	0.0023	0.0023	0.022	0
Hydrogen Cyanide (HCN)	0	16.5	16.5	0	4.3	12.2	0.4	0.4	3.9	0
Sulfuric Acid (H_2SO_4)	0	0	0	0	26	0	0	0	0	0
Ammonium Sulfate (($NH4)_2SO_4$)	0	0	0	0	30	0	0	0	65	0
Total	49085	49094.6	49094.6	17500	15299.7	51343.1	4970.00	5378.89	14937.7	408.909
Temperature (°C)	266.5	400	352	30	54.2	48.6	-65	53.9	49.7	139.6
Pressure (lb/in² absolute)	75	31	29	25	18	18	9.5	14.7	14.7	10
Molecular Weight (g/g-mole)	28.9	28.9	28.9	18	18.4	27.9	77.3	78.8	18.1	104.3
Density (lb/ft³)	0.21	0.069	0.069	62.9	61.8	0.081	60.2	53.7	61.9	64.9

Mass Balances Continued

Intermediates Manufacture Mass Balance (continued)

Stream Number	A24	A25	A40	A41
Phase	Liquid	Liquid	Vapor	Liquid
Component	**Mass Flow (lb/hr)**			
Benzene (C_6H_6)	0.0089	0	33.4	0
Nitrogen (N_2)	0	0	0.0024	0
Oxygen (O_2)	0	0	0.00018	0
Water (H_2O)	0	0	1500	16331
REACtant	199.3	1	1.7	0.0009
PRODuct	1	205.6	0.054	1
IMPURity	1.9	0.0016	0.013	0
Ammonia (NH_3)	0	0	7.8	0.03
Nitric Oxide (NO_2)	0	0	2.1	0.004
Sulfur Dioxide (SO_2)	0	0	2.6	0.002
Carbonyl Sulfide (COS)	0	0	0.0016	0
Carbon Dioxide (CO_2)	0	0	0.022	0
Hydrogen Cyanide (HCN)	0	0	3.5	0.34
Sulfuric Acid (H_2SO_4)	0	0	0	0
Ammonium Sulfate (($NH4)_2SO_4$)	0	0	0	65
Total	202.209	206.602	1551.19	16397.4
Temperature (°C)	31.7	136.3	105.4	108.9
Pressure (lb/in^2 absolute)	0.5	0.65	18	20
Molecular Weight (g/g-mole)	99	109.9	18.4	18
Density (lb/ft^3).	66.1	69.7	0.046	58.9

Process Chemistry Form

Reactor Designation: Fixed-bed Catalytic Reactor

Chemical Reactions:

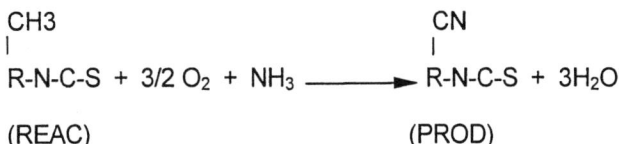

$$\overset{\displaystyle CH3}{\underset{\displaystyle R\text{-}N\text{-}C\text{-}S}{|}} + 3/2\ O_2 + NH_3 \longrightarrow \overset{\displaystyle CN}{\underset{\displaystyle R\text{-}N\text{-}C\text{-}S}{|}} + 3H_2O$$

(REAC)   (PROD)

Unfortunately, at high temperatures (greater than 360°C), the compounds REAC and PROD are thermally decomposed (i.e., oxidized) to CO_2, NO_x, SO_2, COS, HCN, etc

Key Reaction Parameters:

- Inlet to the catalyst bed is about 340°C and 75 psia.

- The reacted gases exit at greater than 400°C and 51 psia.

Order of Feed Material Addition:

REAC, PROD, steam, NH3 and air are mixed and heated prior to entering the reactor.

Order of Product(s) Removal:

The products, byproducts, unreacted feeds exit in one gas stream.

Solvents, Catalysts and Trace Materials:

A fixed-bed catalytic reactor is used. The catalyst is an alumina honeycomb substrate coated with a precious metal catalyst.

Byproduct(s) or Waste(s):

Some REAC and PROD are converted to undesirable byproduct gases such as SO_2, COS, CO_2, NO_x and HCN

Other Pertinent Information to Describe the Reactions and Reactor:

Hot spots occur in the reactor

Component Information

Purpose

Using component information combined with the Waste Stream and Process Analyses a creative person is able to identify process improvements to reduce waste generation.

The minimum information for each component is:

- Name

- Formula

- Molecular weight

- Density

- Normal boiling point

- Normal freezing point

Special information such as solubility factors, activity coefficients, vapor pressures, etc. will be needed to a definitive analysis of the best ideas and is normally done off-line after the initial ranking.

Component Property Form

Name	Formula	Mol. Wt.	Density	T_b	T_s
Products			Lbs./ Cu.ft.	°C	°C
PROD	CN | R-N-C-S	110	55	235	60
Intermediates					
Feed Materials					
REAC	CH3 | R-N-C-S	99	48	132	-30
Ammonia	NH3	17	0.05	-33.4	-77.7
Water	H2O	18	62.4	100	0
Oxygen	O2	32	0.08	-183	-218
Others					
IMPUR	Unknown	85	45	116	-20
Nitrogen	N2	28	0.06	-147	-210
Benzene	C6H6	78	54.8	80	5.5
Hydrogen cyanide	HCN	27	0.06	26	-14
Carbonyl Sulfide	COS	60	0.17	50	-138
Nitric oxide	NO2	46	0.12	21.3	-9.3
Sulfur dioxide	SO2	64	0.18	-10	-76

Density in Lbs./Cu. Ft.
T_b Normal boiling point °C
T_s Normal freezing point °C

Review Chapters

The following chapters from the book "Pollution Prevention: Methodology, Technologies and Practices" were attached to the data package sent to the brainstorming participants.

- Chapter 10: Reactor Design and Operation

- Chapter 11: Use of Water as a Solvent and Heat Transfer Fluid

- Chapter 17 Separation Technology Selection

Index